D0865379

SCIENCE 300 CROSSWORD PUZZLES

Marcel Danesi, Ph.D.

chartwell
books

INTRODUCTION

The first crosswords appeared in England during the 19th century, initially created for children. But in the United States, the crossword soon developed into a serious adult pastime. The first known published crossword appeared in the New York World newspaper in 1913 and was created by Arthur Wynne, a journalist from Liverpool, who since then has been credited as the inventor of the popular puzzle. During the early 1920s, other newspapers started running crosswords and within a decade the puzzles were featured in almost all American newspapers.

Many aspects of modern life are impacted by scientific knowledge. Science is a system for exploring and for innovation, and it responds to societal needs and global challenges. It has a specific role, as well as a variety of functions for the benefit of our society: creating new knowledge, improving education, and increasing the quality of our lives.

If you fancy yourself an amateur scientist, this puzzle collection is for you. It consists of 300 crossword puzzles covering all the main themes related to science, from medicine and math to technology and contemporary ideas.

The puzzles have been organized into three levels of difficulty and can be easily identified by their colors—teal for easy, yellow for moderate, and pink for difficult.

PART 1: EASY

The puzzles in this section will deal with general aspects of science, including scientists and their ideas. To start things off nice and easy, the definitions in this section are as straightforward and as concrete as possible. However, this crossword format involves both letters that cross twice and those that do not, making them a little trickier than the average crossword.

PART 2: MODERATE

These crosswords will deal with aspects of the physical and life sciences. As in part 1, the crossword format involves both letters that cross twice and those that do not. Again, the clues are fairly straightforward.

PART 3: DIFFICULT

The puzzles here will deal with aspects of computers, technology, artificial intelligence, and other modern themes of science. As in the previous two parts, the crossword format involves both letters that cross twice and those that do not. While each puzzle revolves around a theme hinted at in the title, you'll have to rely heavily on your logic to get these solved.

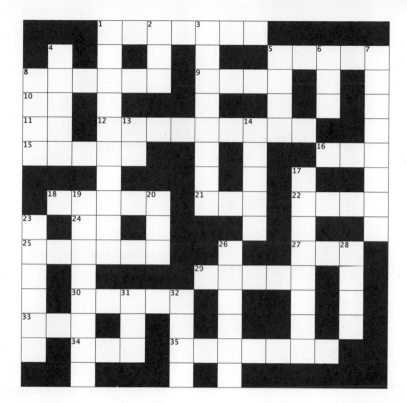

ACROSS

1. Scientist Louis, who discovered how to eliminate pathogens from milk, fruit juices
5. Swiss city where the mathematician Leonhard Euler was born
8. Scientist Sir Isaac, who developed the calculus and the field of optics
9. Appellation
10. Hosp. area
11. Morning hrs.
12. Greek philosopher who wrote the first book on physics; pupil of Plato
15. Light coloration
16. Age
18. Scientist Marie, pioneer in radioactivity
21. Nothing
22. Roman poet who prefigured the modern-day scientific idea of "flow"
24. Sodium *(abbr.)*
25. Inventor of the light bulb
27. Unit of energy
29. Prize for a great achievement in science
30. Poet Dickinson, who was fascinated by science
33. Organ with a hammer
34. Center of the solar system
35. Italian scientist, condemned by the Church for heresy

DOWN

1. Ancient Greek mathematician, whose theorem about right triangles is still studied in schools today
2. Male offspring
3. Scientist Albert, known for the theory of relativity
4. Scientist Enrico, who helped develop the atomic bomb
5. Scientist A. Graham, who gave the first demonstration of the telephone in 1876
6. Oceanography focus
7. Scientist, painter, and inventor, Da Vinci
8. Tidy
13. Symbol for rhenium
14. Scientist Nikola, who invented the first alternating current motor
17. Mathematician Ada, who contributed to the design of modern computers in the 1800s
19. Cosmos
20. Major division of geological time
23. Astronomer Johannes, who discovered laws that govern orbital motion
26. Technologies for handheld computers
28. Audacity
31. Atom or molecule with a net electrical charge
32. Hindu physical and spiritual discipline

Answer on page 303

ACROSS
1. Scientific investigation
7. Correct
8. Electrical _____, which can be positive or negative, is carried by subatomic particle
11. Product that might be found in the kitchen, bathroom, or garage

13. Smallest unit of a chemical compound
14. Greek "H"
15. Boiling
16. It can be nuclear, solar, wind, etc.
17. Pliable (especially in reference to a metal)
19. Nuclear
21. Global

DOWN
1. Mathematical formulas
2. Electron, neutron, or photon, for example
3. Aluminum, copper, and iron, for instance
4. Application of science in a broad way

5. The tooth of a gear
6. Test site
9. What scientists seek
10. Charged
12. Disturbance transferring energy through matter or space
13. Tiny prefix
17. Raw information
18. Existence
20. Planet orbiter

Answer on page 303

ACROSS
3. Precise
5. The theory of ____, one of Einstein's; greatest achievements
7. Data, real information
10. Phase
11. Cambridge univ.
12. Food regimen
13. Symbol
15. Self-evident principles
17. Negative response
18. Branch of physics
19. Representation, such as Rutherford's " ____ of the atom"

DOWN
1. Discipline that includes arithmetic, algebra, geometry, and trigonometry
2. Darwin's theory
4. Kind of tube found commonly in science labs
6. Discipline studying the distribution of wealth
8. Supposition
9. Concepts
14. Darwin's "On the ____ of Species"
15. A building block of all matter
16. Common rodent

Answer on page 303

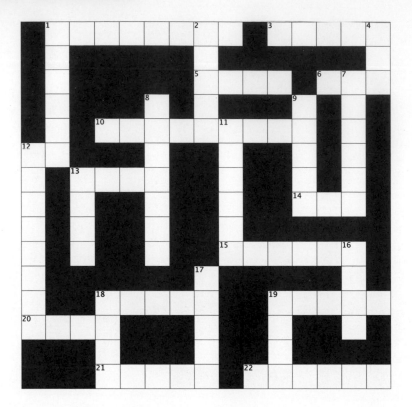

ACROSS

1. Astronomer's field
3. "The final frontier"
5. Region
6. Era
10. Device that allows us to see faraway stars
12. Thus
13. Earth's natural satellite
14. Home of the stars
15. Orbiting laboratory that was launched in 1973 and abandoned in 1974
18. Like some eclipses
19. Like some eclipses
20. Terrible fate and name of mountain range on Saturn's largest moon
21. Descriptor for some stars
22. Shooting star

DOWN

1. Seventh planet from the Sun
2. Luminous celestial bodies
4. Night before
7. The Milky Way, for example
8. Celestial bodies that orbit the Sun
9. Planet named after the roman goddess of love
11. Complete and organized system
12. Small rock orbiting the Sun
13. The red planet
16. "Ursa" in Ursa Minor
17. Earth's journey around the Sun
18. Polluted foggy air
19. Existence

Answer on page 303

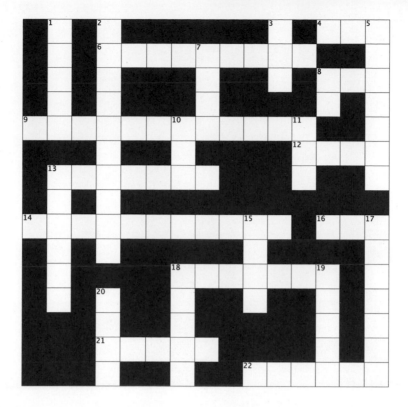

ACROSS

4. Provider of heat to the Earth
6. Discipline studying human behavior
8. Armed bellicose conflict
9. Discipline studying humanity
12. Opposite of under
13. Complex of languages, symbols, arts, rites, etc. that characterize human groupings
14. The science of language
16. Age
18. Groupings of people according to wealth
21. Social conventions, standards
22. Organisms

DOWN

1. Our species
2. Science of the mind
3. Past
5. Natures opposite in a question of human development
7. Aware of
8. Royal pronoun
10. Equal value
11. A common personal pronoun
13. State of emergency
15. Type of study focusing on a specific individual or group
17. Living organisms, different from plants
18. Subject for forensic scientists
19. Country that occupies the greater part of the Iberian Peninsula
20. Clique

Answer on page 303

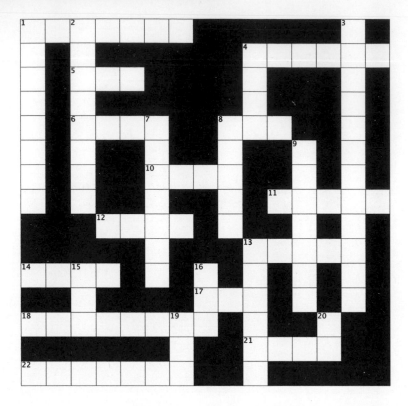

ACROSS

1. Arithmetic that uses letters
4. Name of the number system consisting only of 0's and 1's
5. Any whole number *not* divisible by "2"
6. Whole numbers divisible by "2"
8. A number collection
10. Prefix meaning one
11. A whole number divisible only by "1" and itself
12. Although it means "nothing," it is used in positional number systems as a "place holder"
13. Number
14. Average
17. End of a professor's email address
18. Number expressing the power to which a number is to be raised
21. Systematic thinking
22. What the system made of ten numbers is called

DOWN

1. The opposite of subtraction
2. Branch dealing with triangles, circles, and squares
3. Branch concerned with triangles
4. The "5" in "53"
7. Figure
8. Number of points
9. Ideal, perfect example
13. Twofold
15. Phone purchase
16. Take-home
19. Nothing
20. A ratio that applies to the circle, equal to 3.14…

Answer on page 303

ACROSS

1. Building
7. Broad blade
8. Balance
10. Elevated railroad (*abbr.*)
11. Placed
13. Invent
15. Unmovable, often contrasted with *dynamic*
17. Movement of liquid because of pressure
18. It falls as water droplets
19. Reflexive pronoun
20. Rigid bar with a fulcrum, used to lift heavy objects

DOWN

1. Kind of engineer who designs infrastructures (roads, bridges, etc.)
2. Kind of engineer who deals with the design and operation of marine vessels and structures
3. Kind of strength
4. Element or compound
5. Mount
6. Kind of engineer who designs power producing devices
9. Region
10. Applied science concerned with the design and operation of structures
12. Kind of engineer who deals with certain systems
14. Now
15. Examine closely
16. Wrestle
19. Famed Cambridge Univ.

Answer on page 304

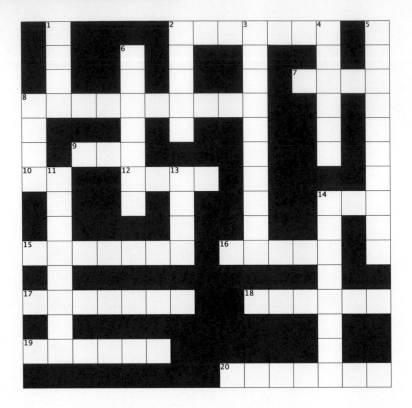

ACROSS

2. The scientific method, for one
7. Implausible account
8. Untruths
9. Tug
9. Sch. famous for scientific and technical research, located in Cambridge, Massachusetts
10. Trauma ctr.
12. Impression
14. Number of digits in the decimal system
15. Describes accurately
16. Science _____, ideas that are not true
17. Forecast
18. Someone who takes the ideas of someone else as their own
19. Coincidence
20. Ideas that are accepted as such, without scientific proof

DOWN

1. Collected facts
2. Evidence supporting a theory
3. Theory that an unexplained event is caused by a covert group
4. Unchanging
5. Certainty
6. Doubter
8. Bogus
11. Systematic scientific investigation
13. Conclusions
14. Like information learned in school

Answer on page 304

ACROSS

4. Physical surroundings
7. Prevailing weather conditions
8. Breathe
9. Condition
10. Global concern of the gradual increase in the temperature of the Earth's atmosphere
12. It blows
14. Mild
16. Type of energy harnessed from the Sun
17. "Carbon _____," amount of greenhouse gases generated by our actions

DOWN

1. It is called the "zone of life" on Earth
2. Glacial
3. Marsh or swamp
4. "Wind _____," power generated from the wind
5. Our planet
6. Atmospheric conditions
10. Capital
11. Direct effect
12. As
13. Inhabit
15. Back

Answer on page 304

ACROSS

3. Galileo designed a mechanism for the pendulum _____ (time-keeping device)
5. Galileo's nationality
7. Creative activity, such as music or painting
8. Galileo used the telescope to observe the moons of this planet, the largest in the solar system
9. Galileo made a topographical _____ (geographical diagram), estimating the heights of a mountain
10. With his telescope, Galileo was able to give an in-depth description of the Earth-orbiting satellite
14. Crime Galileo was tried by the Inquisition for
15. Opposite of minus
16. Alike
17. Galileo was among the first to graph the phases of this planet, the second planet from the Sun
18. Celestial bodies

DOWN

1. City of the leaning tower, where Galileo was born
2. Galileo was the first to describe the rings of this planet, the sixth from the Sun, in an in-depth fashion
3. Galileo's championing of Copernican heliocentrism met with stern opposition from within the Catholic _____, leading to his Inquisition
4. Galileo promoted the heliocentric model of this Polish astronomer
6. Famed sch. In Cambridge
11. Italy's capital, where Galileo went to defend himself
12. Galileo is sometimes called the "father" of this science of matter
13. Galileo _____
14. Underworld
15. Celestial bodies

Answer on page 304

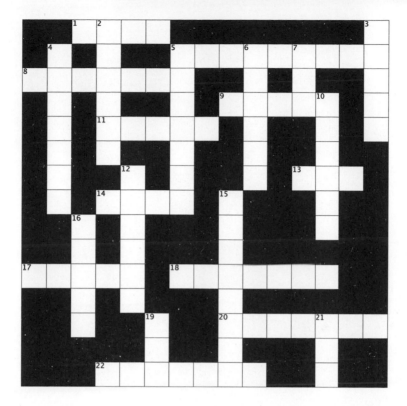

ACROSS

1. Objective
5. Newton's profession at Cambridge
7. Newton's nationality
8. Newton employed this object, which refracts light, for his research on colors
11. Reason
12. For his accomplishments, Newton was knighted with this title
14. Newton's _____ of motion became the foundation of physics at the time
17. Newton was inspired to develop his theory of gravity by watching this fruit fall from a tree
18. Newton was intrigued by this craft, the forerunner of chemistry
19. German mathematician with whom he shares credit for the invention of the calculus
22. Method

DOWN

2. The science of light, to which Newton made key contributions
3. Number that has no divisors other than itself and "1"
4. Newton's work on this force, which keeps objects from flying off the Earth, laid the foundations for modern physics to develop
5. Branch of science for Newton
6. Cultivator
7. Bro's sibling
10. Newton's laws of _____ (movement) remain central to physics
12. Physics is, overall, the science of *this*
15. Branch of mathematics that he invented, simultaneous to Leibniz in Germany
16. Use
19. Number of digits in the binary system
20. Zero

Answer on page 304

ACROSS

2. Einstein accepted a fellowship at the Institute of this Ivy League university in 1933
4. The "E" in $E = mc^2$— Einstein's most well-known formula
5. Zero
9. Area for which Einstein is famous
11. While married, Einstein fell in love with Elsa Löwenthal, who was this kin relationship to him
12. Einstein's adoptive country *(abbr.)*
14. Country of Einstein's birth
16. String instrument played by Einstein
18. Scientific discipline reshaped by Einstein
21. Loosen

DOWN

1. Einstein is known broadly for his theory of _____
3. Sick
6. Pancreatic hormone
7. Einstein made key contributions to _____ mechanics,"
8. The "c" in $E = mc^2$
10. The "m" in $E = mc^2$
11. Assertion
13. Einstein contributed to our knowledge of this force, which keeps objects from floating away from Earth
15. Physical movement
16. Worth
17. Einstein was awarded this Swedish prize for physics in 1921
18. Univ. teacher
19. Ivy League university in New Haven
20. Only

Answer on page 304

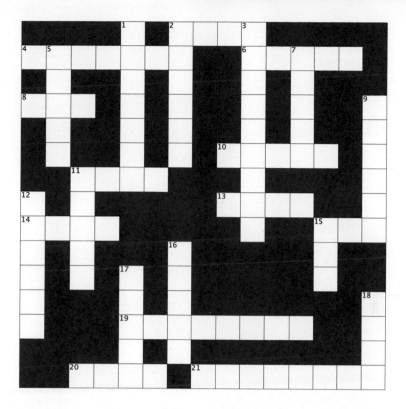

ACROSS

2. Small reddish planet, fourth from the Sun
4. Planet that is eighth from the Sun; named after the Roman god of the sea
6. Basic units of matter
8. Age
10. Small planetary body orbiting the Sun; named after the Greek god of the underworld
11. The hottest part of the Sun
13. Luminous celestial body
14. Objectives
15. Home of the stars
19. Opposite of acidic
20. Existence
21. Small astronomical body orbiting the Sun

DOWN

1. Largest planet of the solar system
2. Falling star
3. The moon, for example
5. Home planet
7. Path of a planet around the Sun
9. The Milky Way, for example
11. Celestial body consisting mainly of ice and dust, which burns up as it enters Earth's atmosphere
12. Sixth planet from the Sun, named after the Roman god of agriculture
15. Perceive
16. "_____ Way," the galaxy that contains the solar system
17. Descriptor for a star that is small in size and low in luminosity
18. Very dry

Answer on page 305

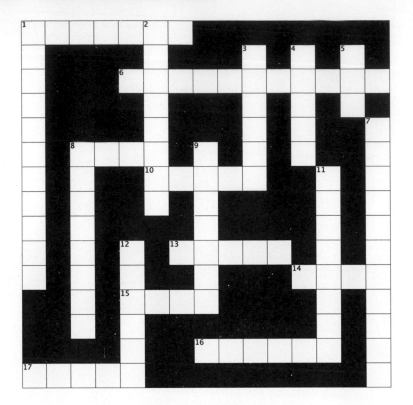

ACROSS

1. Official who investigates suspicious deaths
6. Unique impression made on a surface
8. Flesh
10. Pieces of evidence used in crime detection
13. Focus of forensic dentistry
14. Mark left by a wound that has healed
15. Goals
16. Cadaver
17. Pieces of forensic evidence, especially from the head

DOWN

1. The study of deviant behavior
2. Collection of clues
3. Rap sheet entries
4. Tiny amount
5. Type of evidence that can help identify a person by their genes
7. Person who committed a crime
8. "Blood ____," patterns of blood stains at a crime scene
9. Projectiles
11. Murder
12. Impressions left by tires

Answer on page 305

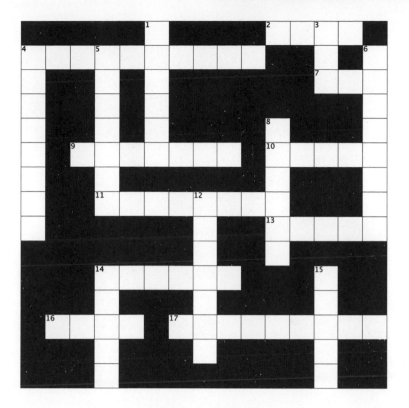

ACROSS
3. Sense
6. An ounce is worth a pound of cure, in a common saying
8. Nail site
10. Medical disorder
11. Give medical care
12. Physical activity, which is highly recommended to promote health
14. Protects
15. Protection, care
17. Nourishment
18. Doctor's order

DOWN
1. Withstand
2. Food regime
4. Consume food
5. Widespread disease outbreak
6. Immunizing shots
8. The study of hereditary traits
9. Pressure, worry
13. Part of the CDC
15. Camera used by some doctors
16. Flourishing

Answer on page 305

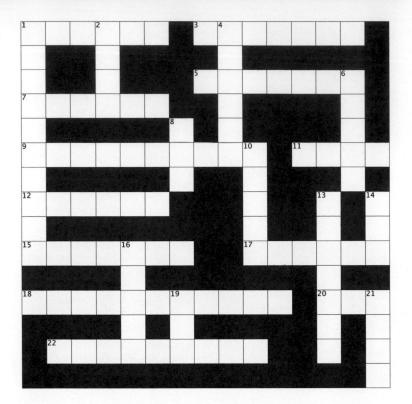

ACROSS
1. Of the stars
3. Renaissance Italian astronomer, who first observed craters on the moon
5. Space mission program launched by NASA in 1977
7. Spacecraft
9. Course-plotting
11. Star _____, TV series and set of movies about the exploration of space
12. Seventh planet from the Sun
15. Soviet artificial satellite, launched in 1957
16. Planet circled by flat broad rings
17. Building with an astronomical telescope
19. Male turkey
21. Astronomer's optical instrument

DOWN
1. Space explorers
2. Boulder
3. American space program that was the first to land astronauts on the moon in 1969; named after the son of Zeus
6. Space vehicle controlled remotely
8. Greek "H"
10. Stars showing a sudden large increase in brightness
13. Spacecraft able to land
14. "One small step for a_____; one giant leap for mankind;" Neil Armstrong's statement when he stepped on the moon in 1969
16. Like Odin or Thor
19. Part of a circle
21. Earth's natural satellite

Answer on page 305

ACROSS

1. Large vessel
3. Cruise
7. It complements latitude to plot location
8. Ferry
9. Kind of admiral
10. Speed unit equivalent to one nautical mile per hour
12. Line around the center of the Earth
16. Sea _____ : a sailor
18. Instrument with a magnetized pointer for determining direction
20. Line of longitude passing through a given place and the terrestrial poles

DOWN

1. Bodies orbiting the Earth that are used for navigation, among other things
2. Steering a ship or an aircraft
3. Star at the center of the solar system
4. It complements longitude to plot location
5. Detection system using radio waves
6. On
11. Ocean contents
13. Small boat used for emergencies
14. Diagrams of territories
15. Steering pole
19. Luminous celestial body

Answer on page 305

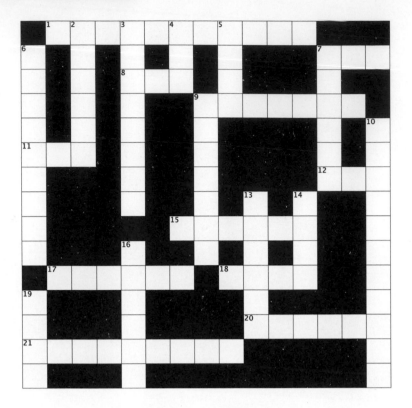

ACROSS
1. The science of maps
7. Part of a circle
8. Distress signal
9. Latvia, Lithuania, and Estonia
11. Initials for a beer with a country in its name
12. Ice _____ , a covering of ice over a broad region
15. Its capital city is Moscow
17. Continent containing France, Spain, Italy, and Germany, among others
18. The red planet
20. Like someone from Copenhagen
21. Landform surrounded by water, while being connected to a mainland

DOWN
2. Second largest continent
3. A state in Australia
4. Navig. tool
5. The largest continent
6. Asia, Africa, North America, South America, Antarctica, Europe, or Australia
7. It's around the North Pole
9. Its capital city is Brussels
10. Kind of map showing natural features of an area
13. Landmass surrounded completely by water
14. Atlas contents
16. Its capital city is Warsaw
19. _____ Cod, peninsula in southeastern Massachusetts

Answer on page 305

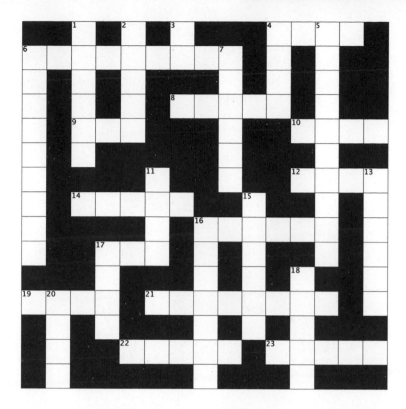

ACROSS

4. Prepare food for consumption
6. Abnormal responses to foods
8. Word after food or before restaurant
9. Finish
10. _____ food, quickly prepared food
12. Staple food for over half the world
14. Food from a tree
16. Chemical symbol is C
17. Fat often used in cooking
19. Beef or pork, for instance
21. Chemical building blocks—amino acids
22. Wheat cultivated as food
23. Breakfast food

DOWN

1. Substance present in grains that might cause illness in some people
2. Widely eaten staple food
3. Ratio of the circumference to the diameter
4. It can be eaten on the cob
5. Procedures
6. Flavor preservers or enhancers
7. Bread, milk, or eggs, for example
11. Tablet
13. Vigor
15. Like a TV dinner
16. Unit of food energy in the form of heat
17. Type of cereals
18. Food that can make the eyes water
20. Consumes food

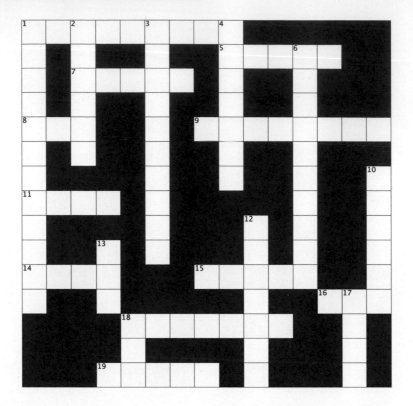

ACROSS

1. Enclosed system supporting life
5. Cultivated plants
7. Lawns, gardens
8. Frozen water
9. Contaminated
11. Actual, not false
14. It comes in waves
15. Opposite of rural
16. Source of solar energy
18. Its change is caused by global warming
19. Expunge

DOWN

1. The variety of life on this world
2. Chemical symbol is O
3. Concerning the properties of water
4. Branch of biology concerned with organisms in their physical surroundings
6. Inhabitants as a whole
10. Chemical element C
12. An animal's natural home
13. Large body of water
17. Puts to work
18. Automobile

Answer on page 306

ACROSS

4. Classification system of species
6. Plants as a category; opposed to fauna
9. Some warm-blooded animals
12. Its capital is Kathmandu
13. Group of living organisms
16. Animals with a backbone
18. Flying animals
20. Hive dwellers
21. Animal on California's flag
22. They live only in water
23. Terrapin

DOWN

1. Animals as a category; as opposed to flora
2. Living organisms classified apart from plants
3. Australian bird resembling the ostrich
5. Living organisms classified apart from animals
7. Snakes and lizards, for example
8. Large animals with tusks and long noses
9. Common rodent
10. Small, burrowing, insect-eating mammal
11. Part of a plant that holds it up
14. Species with common features classified together
15. Animal such as a chimp or gorilla
17. Plants with a trunk and branches
19. Life-nourishing precipitation

Answer on page 306

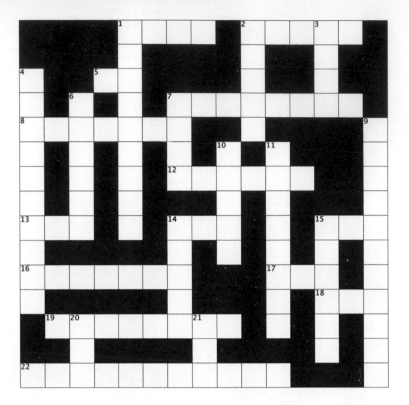

ACROSS
1. Facts, information (especially numerical)
2. A specific scientific investigation
5. Common HS club
7. Reciprocally
8. Statistical mean
12. Standard
13. Number of digits in the decimal system
14. Deoxyribonucleic acid
15. Title for Isaac Newton
16. Gather
17. Rod
18. Animal doc
19. Breakdown (of findings)
22. Statistical relationship

DOWN
1. Standard _____, refers to the degree to which something strays from the average
2. Comb through
3. Opposite of interesting
4. Branch of mathematics for analyzing numerical data
6. Mid value in a frequency distribution
7. Statistical average
9. Facts in general
10. Experiment conducted to see if something is suitable
11. It assumes different values in an experiment
14. Times
15. Study that may involve interviews
20. Neither's partner
21. Bank statement abbr.

Answer on page 306

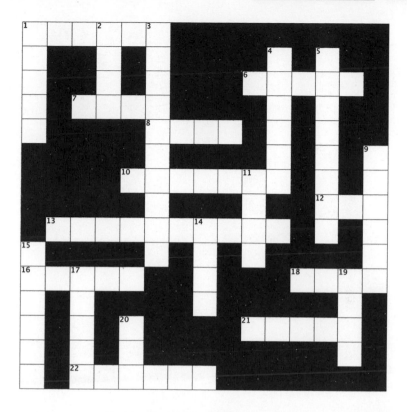

ACROSS

1. Measurement system used by most of the world
6. Odometer units
7. Each one is 12 inches
8. One-twelfth of a foot
10. Temperature scale on which water freezes at 00
12. Unit of energy
13. Temperature scale on which water freezes at 320
16. Enter
18. It is measured, in hours, minutes, days, etc.
21. Units of electromotive force
22. Ten years

DOWN

1. Basic unit of measure, often used in race distances
2. Measure of constant change
3. Small metric measure
4. Metric units of capacity
5. Metric units of area
9. What the superscript "°" stands for in, say, 12°
11. Basic measure, such as meter
14. Sixty minutes
15. Sixty seconds
17. Unit of weight
20. Small unit of time, for short

Answer on page 306

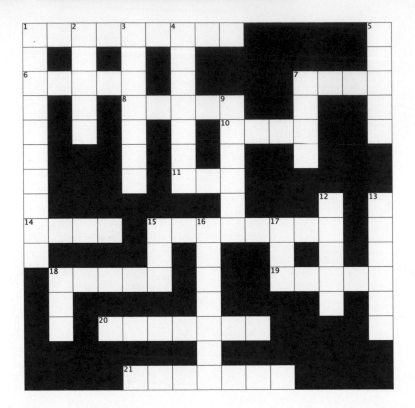

ACROSS

1. Historically valuable objects
6. Agricultural yields
7. Water-carrying containers
8. Places studied archeologically
10. Crypt
11. Age
14. Willing to play, or what you might play
15. Dig, archeologically
18. Ossified substances
19. Basic type of community
20. Study of the past
21. Family or blood ties

DOWN

1. Science of human history and prehistory
2. Instruments made by homo sapiens as far back as the Stone Age
3. Remnants
4. Values, rites, etc. shared by a social group
5. Urns
7. Occupations
9. Rock layers
13. Spiritual symbols sometimes found on poles
15. Suffix with count or host
16. Traditions specific to a society
17. Painting, sculpting, or dancing, for example
18. Dates before A.D.

Answer on page 306

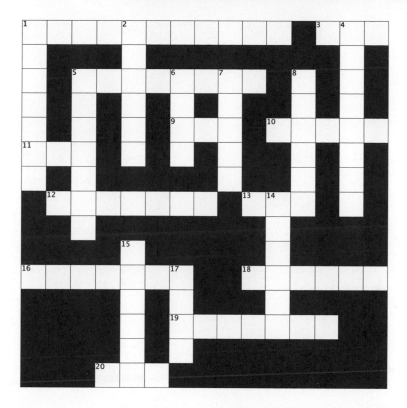

ACROSS

1. Discipline to which Archimedes contributed most
3. Time period
5. City of Archimedes' birth
9. What we breathe
10. Lifting bar with a fulcrum, explained by Archimedes
11. Head gesture
12. Military arms, which Archimedes designed
13. Night before
16. Archimedes discovered how to calculate *this*, while taking a bath
18. Wheel on an axle designed to support movement of a cable or belt, which Archimedes studied
19. What Archimedes wrote a famous treatise on; they are used for counting (among other things)
20. Directed

DOWN

1. Archimedes helped develop the concept of *this* (apparatus applying mechanical power)
2. Exclamation that Archimedes shouted after his discovery while taking a bath
4. Branch of mathematics to which Archimedes contributed in a significant way
5. Legend has it that Archimedes was killed by a ____; that is, someone who serves in the army
6. Talon
7. Archimedes also invented a ____, for transferring water into irrigation ditches
8. Archimedes' country of birth
14. Capacity (of water, for example)
15. Figure of 360°, which Aristotle studied (determining the ratio π to great precision)
17. Tug

ACROSS

2. Opposite of soft
4. Mathematical study of the structure of random events
8. Self-importance
9. Controlled scientific studies
10. Study intended to indicate the viability of an investigative method
12. Lab balance
13. Scientific explanation
16. Research, investigation
19. Take the size, amount, etc. of something

DOWN

1. Inventory
2. Scientific supposition
3. Infer systematically
5. Systematic investigation
6. Glitch
7. Conclusion based on given facts
11. Trying out
13. Arboreal plant
14. Explore with an instrument
15. Not quite right
17. Opposite of far
18. Creative activity such as painting

Answer on page 307

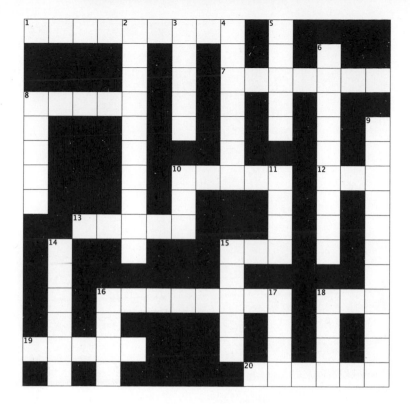

ACROSS

1. Figures made up of three sides and angles
7. "_____ of the spheres," an expression used by Pythagoras in reference to the musical sounds made by heavenly bodies as they moved
8. Any number divisible only by "1" and itself
10. Art that Pythagoras studied for its harmonic structure
12. Portion
13. Opposite of dark
15. Decay
16. Branch of mathematics to which Pythagoras made significant contributions
18. Any creative activity
19. Mathematical demonstration
20. Dangerous predicament

DOWN

2. Belief in the mystical properties of numbers— founded by Pythagoras
3. Connections (such as those between numbers and reality)
4. Solid circular objects, such as the Earth
5. Our planet
6. Science of the origin and development of the universe
8. Greek philosopher influenced deeply by Pythagoras
9. Pythagoras is sometimes called the "father" of this
10. Famed sch. In Cambridge
11. Type of sect, used to describe the Pythagoreans
14. An explanation in science or mathematics
15. π, for example
16. Shine
17. January – December
18. Objectives

Answer on page 307

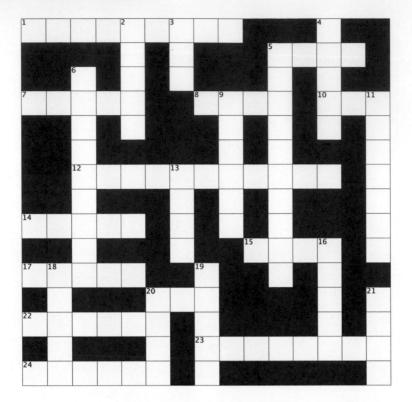

ACROSS

1. Alexander Graham Bell invention
5. Medicinal tablet
7. One of four, allowing a vehicle to move
8. Noble gas in some bright lights
10. Automobile
12. Discovery by happenstance
14. Handheld instruments
15. It generates smoke
17. The printing _____, which made possible the mass production of written materials
20. Mathematician Lovelace
22. Inventor Thomas
23. Computer network
24. Coloring pastel

DOWN

2. Canvasses
3. Boat blade
4. Timepiece
5. An antibiotic produced by certain blue molds, discovered in 1928 by Scottish scientist Alexander Fleming
6. Astronomer's optical instrument
9. Motor
11. Missiles
13. Hammer's aim
16. Kind of salt
18. Tool used in aviation and navigation to determine location
19. Wireless communication system invented by Guglielmo Marconi
20. Unknown auth.
21. Ballot

Answer on page 307

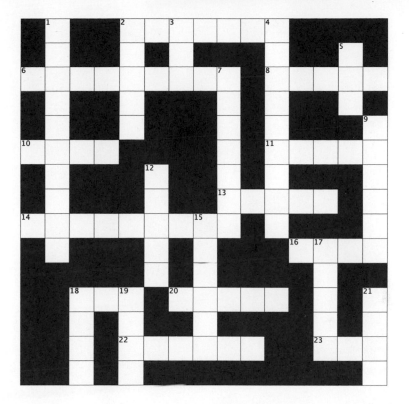

ACROSS

2. Seals, washers
6. Working parts of machines
8. Propeller blade
10. Latch
11. Like some waves
13. Narrow valley or mass of ice obstructing a narrow passage
14. Clasps
16. High value playing cards
18. Maritime org.
20. They are hammered
22. Twisty hardware items
23. Nothing

DOWN

1. Temperature-regulating device
2. Mechanisms
3. Hit the slopes
4. Arrangement
5. Gear
7. Metal coils
9. Pincers with flat surfaces used for gripping
12. Ladder steps
15. Spin
17. Series of metal rings connected together
18. Standard
19. Metal tool used to hold an object firmly in place
21. It fastens a door or window

Answer on page 307

ACROSS

2. American space agcy.
4. Home of astral bodies
6. Meteorological conditions
7. The science of plants
9. May honoree
12. The science of animals
14. Sick
15. The science of matter
18. Consume food
19. Mediator, briefly
21. The science of the past
23. The science of the stars and outer space
24. Uninvited picnic guest

DOWN

1. Prefix meaning "of the stars"
2. Opposite of old
3. Having magnitude, not direction; contrasted with *vector*
5. Science class with the periodic table of elements
8. Prefix meaning "new"
10. The discipline of Hippocrates
11. The science of rocks and related phenomena
13. One of three states of matter, the other two being solids and liquids
15. Medicinal tablet
16. Natural habitat of many aquatic mammals
17. The science of living organisms
18. Organ with a hammer
20. Lower bodily extremities
22. Automobile

Answer on page 307

ACROSS

1. Madame Curie's birth nationality
5. Curie was awarded this prize in physics in 1903 and in chemistry in 1911
7. Curie's work contributed to treating this deadly disease
8. Subject that Madame Curie was a pioneer in
9. Group working together
11. Madame Curie's first name
14. Female offspring
17. Madame Curie's profession
18. Bigotry faced by Madame Curie
20. French city where Madame Curie carried out her work

DOWN

1. Chemical element with the symbol Po, discovered by Madame Curie and her husband in 1898
2. Name of Madame Curie's husband
3. Chemical element with the symbol Ra, discovered by Madame Curie and her husband along with polonium
4. Rebuked
6. Age
10. Shrewd
12. Beams of light or radiation
13. Stresses
15. City of Madame Curie's birth
19. Prefix with center

Answer on page 308

ACROSS

2. Goes around, as celestial objects do
5. Chemical element C
7. Frozen water
8. The Alps or the Rockies, for example
10. American space agcy.
12. Era, historical time frame
14. Demand
15. Fertile loamy soil
17. Wetlands
20. Etna or Vesuvius, for example

DOWN

1. Earth oldest living ecosystem, with a large canopy
2. Happening
3. Bodies of water such as the Atlantic or the Pacific
5. Protective layer around the Earth
6. Organic matter used as a fuel
7. Map line connecting points with the same temperature
9. Huge waves influenced by the moon
11. Living organisms classified separately from plants
13. Bat's home
16. Esoteric
18. Earth's atmosphere
19. Protect

Answer on page 308

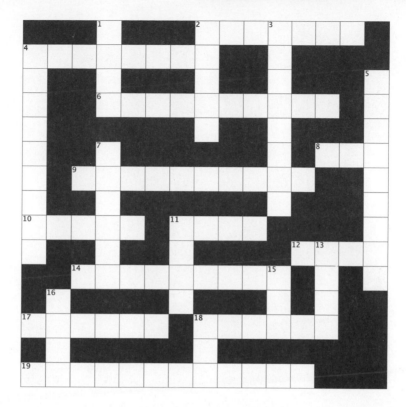

ACROSS

2. Craft aiming to turn metals into gold
4. Cure-all solutions
6. Study of the occult meaning of numbers
8. Goal
9. Now defunct discipline which believed that the shape and size of the cranium indicated character and mental abilities
10. Distress signal
11. Story invented as an explanation of something unknown
12. Water-carrying vessels
14. Study of celestial bodies as having influence on human life
17. Chart of astrological signs
18. Mystical, mysterious
19. Viewpoints

DOWN

1. Caution
2. Prefix denoting the stars
3. Having the same structure; for instance, the wings of bats and birds, the arms of primates, the legs of canines, etc.
4. Like the motion of a machine that would run forever
5. Animal magnetism
7. Talismans
11. It is visible at night, and has been associated with occultism since antiquity
13. Tons
16. The center of the Earth
18. Not at home

ACROSS

1. The study of space and shapes
5. It is equal to 3.14…
6. Space between two lines which meet at a vertex
8. Two-dimensional space
9. Facts and figures
12. Self-evident notions in mathematics
13. Geometrical propositions proven by logic to be consistent or true
15. Mathematical demonstrations
16. Particular shape; for example, triangle, square, etc.
18. It is 180°
19. Euclid established a school in this Egyptian port

DOWN

2. Title of Euclid's famous work, consisting of 13 books
3. Euclid's work is considered the first textbook in this discipline
4. Accurate
5. They have location
7. Euclid is sometimes called the "_____ of geometry"
10. Euclid's approach provided one of the first models of *this*
11. Suppose something to be the case
14. Observe something in detail
15. Usual
16. A number divisible only by itself and "1"; Euclid proved that these numbers formed an infinite set

Answer on page 308

ACROSS

6. Process of speeding up

7. Demonstration of something as true

8. Approach

10. It is measured in minutes, hours, days, years, etc.

11. Quantity with direction and magnitude

12. Base of common logarithms

14. What the "R" in "D = R ′ T" stands for

18. Chart type

19. Distance

20. Rotate

21. Covers

22. Highest part of a wave

DOWN

1. "_____ speed," extremely high speed—a term popularized by *Star Trek*

2. Speed

3. Branch of mathematics that studies change—invented simultaneously by Newton and Leibniz

4. The product of mass and velocity

5. Process of slowing down

9. Power, energy

11. Empty

12. Giant

13. Propulsion

15. System of measurement used by most of the world

16. Opposite of fast

17. Musical rate of speed

19. Go in a direction

Answer on page 308

ACROSS

3. Sixty minutes
6. Era
7. One sixtieth of an hour
8. The personification of time in Greek philosophy
10. Perceive
11. Nothing
12. Previous time period
13. Epoch
15. Moon goddess
17. One sixtieth of a minute
18. After the present
20. The science of time measurement

DOWN

1. Instant
2. Timepieces
4. Type of energy that comes from the sun
5. Musical time
6. Current time period
8. Chart showing the days, weeks, and months of a particular year
9. Void
14. Number of years one has lived
19. Time zone in NYC

Answer on page 308

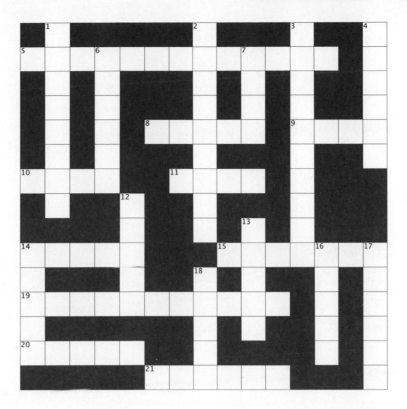

ACROSS

5. Examination
8. Warn
9. House covering
10. Prejudice
11. Colleague
14. Designs
15. Carry out
19. Something witnessed
20. Significant time era
21. Inner mental essence

DOWN

1. Scientific examination
2. Concordance (of ideas)
3. Kind of experiment conducted by minimizing extraneous interferences
4. Ascertain
6. Moral principles
7. Type of tubes found commonly in chemical labs
12. Surgical beam
13. Paint choice
14. Examine in great detail
16. Opposite of lower
17. Tests of something
18. Investigation with a specific goal

Answer on page 309

ACROSS

7. Star orbited by the Earth
9. Variety of life forms in their habitats
11. Epoch
12. Line that deviates from being straight
13. Phases
14. Device that emits an intense beam
17. There are three of these in our world (or so it seems)
20. Transforming a system into an automated one

DOWN

1. Group of people with a specific interest
2. The genetic material of an organism
3. Molecules that do not appear to exist in a stable form in our universe
4. Biological communities
5. A proton or an electron, for example
6. Theory of many worlds comprising everything that exists
8. Program
11. "_____ hole," region of space having an intense gravitational field
15. Change
16. Existence
18. Recognizable figure
20. Natural satellite orbiting Earth

Answer on page 309

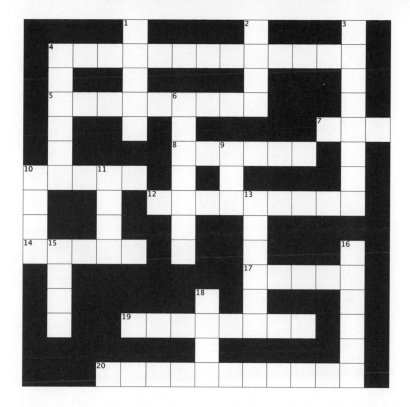

ACROSS

4. Perimeter of the circle

5. The science of the stars, to which Eratosthenes contributed significantly

7. Collection

8. Technique

10. A number divisible only by itself and "1", which Eratosthenes studied

12. Study of the physical features of the Earth, of which Eratosthenes was a founder

14. Ancient city now called Aswan, where Eratosthenes conducted his famous experiment on the circumference of the Earth

17. Delete

19. Developer of a new creation, like Eratosthenes

20. The science of quantity and space

DOWN

1. Tally

2. The city of King Priam, in Homeric legend

3. The study of shapes, figures, angles, etc.

4. Large cavity in the ground

6. Mathematician's interest

9. Yank

10. Authorization

11. Earth's natural satellite

13. Reiterate

15. 12 months

16. Prodigy

18. Roll call answer

Answer on page 309

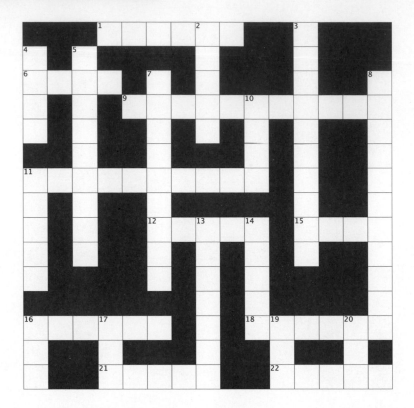

ACROSS

1. Hoist, lifter
8. Type of wind instrument
11. Recorder of heart muscle activity
12. In a car, it shows the speed at which one is going
13. Test ____, found typically in a chemistry lab
15. Writing in verse
16. Bunsen ____ also found in a lab
18. Glass used in science experiments
21. Measuring device
22. Tool that helps pilots determine location

DOWN

2. Ooze
3. Device used to look at organisms not visible to the naked eye
4. Snare
5. Device for looking at the stars
7. Instrument for measuring atmospheric pressure
8. Instrument for measuring temperature
10. Complete
12. Pincers
14. Clean thoroughly
16. Tad
17. Bother
19. Make a mistake
20. Time period

Answer on page 309

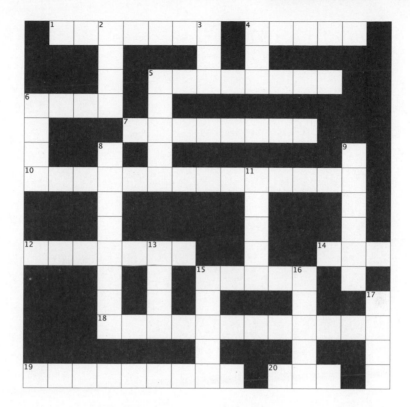

ACROSS

1. Procedures
4. Robot
5. Communicative response, reaction, fed back to the sender
6. Test _____, found commonly in a lab
7. Contraptions
10. The application of very small things
12. Remote _____, device that operates a machine remotely
14. Time frame
15. Exposé
18. Reconnaissance
19. Laptops and desktops
20. Web

DOWN

2. Secure
3. Feminine pronoun
4. Amount from a tube
5. Blaze
8. Branch of technology that deals with automation
9. Part human and part machine
11. Exceed
13. Opposite of under
15. Lance
16. Two-dimensional surface
17. Quiz

Answer on page 309

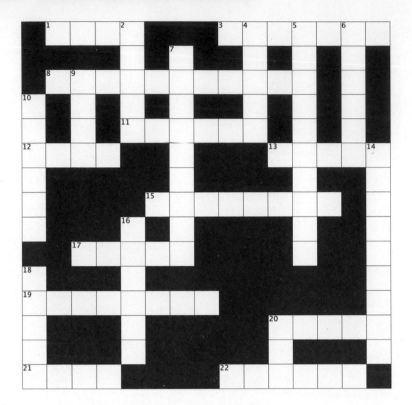

ACROSS

1. Objective
3. Accurate
8. Creative thought
11. Make
12. Information
13. Estimate
15. Scientific examination
17. Detailed investigation
19. Approximate
20. Put forward
21. Clairvoyant
22. Intention

DOWN

2. Rational thinking
3. Lift
5. Culmination
6. State of inactivity
7. Supposition from a set of facts
9. Liquefy
10. Understand
14. Conjecture
16. Strict, following a system
18. Searches for
20. Fairytale vegetable

Answer on page 309

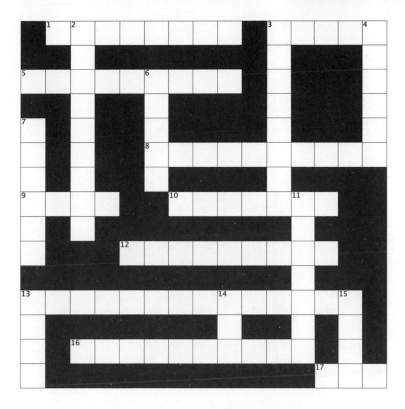

ACROSS
1. Connected systems (for example, of computers)
3. Lesson
5. Community of mutually interacting biological organisms
8. Type of intelligence
9. Pay attention to, as a warning
10. Inhibit
12. Physical organization, construction
13. E-mails, for example
16. Classified items
17. The Mediterranean, for one

DOWN
2. Community of living organisms in conjunction with the nonliving components of their environment
3. Imaginative
4. Gregarious
6. System that includes the Sun and planets
7. Procedure, approach
11. Bodily system that coordinates actions and sensory information
13. Manage
14. The atmosphere
15. Secure

Answer on page 310

SCIENCE AND PHILOSOPHY

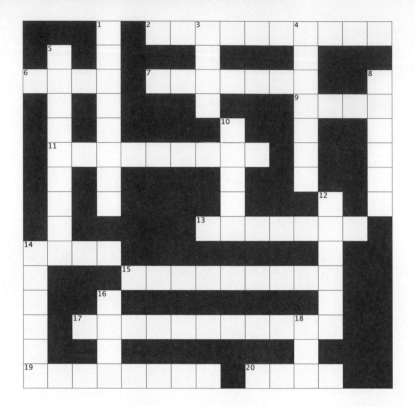

ACROSS

2. Greek philosopher who first came up with the theory of the atom
6. Requirement
7. Opposite of artificial
9. America space agcy.
11. Man who is thought of as the greatest of all Greek philosophers
13. Reasonable
14. Requests
15. China's most well-known philosopher
17. Philosophy dealing with the principles of things
19. Plato's great teacher
20. Philosopher Immanuel, who saw sensation as the basis of knowledge

DOWN

1. Follower of the system of knowledge and philosophy founded by Siddartha Gautama
3. Fable
4. Sickness
5. French philosopher René, who famously said "I think, therefore I am"
8. Mathematician Blaise, who is known for his famous triangle
10. Socrates' student
12. The reasons behind phenomena
14. Associates
16. Terror
18. Able to

Answer on page 310

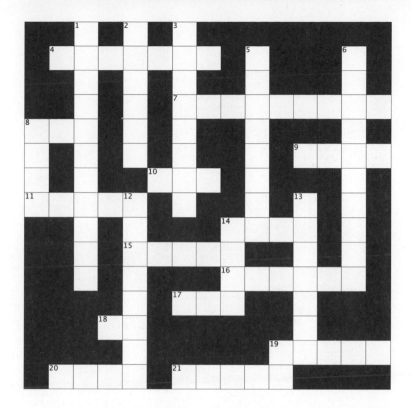

ACROSS

4. Number system consisting of ten digits
7. Numbers formed with Ö-1
8. Math func. that is the inverse of an exponent
9. Collections of mathematical objects
10. Break
11. A number divisible only by itself and one
14. Place-holder in a positional system
15. Connect
16. Regular
17. Number of digits in the binary system
18. π
19. Numerical proportion
20. Like a number divisible by two
21. Arch

DOWN

1. Numeral system with sixty as its base; used commonly in measuring time
2. Numeral system consisting of the digits "1" and "0"
3. Logical
5. A number below zero
6. A number such as Ö2
8. Ring
12. x + 3y = 23, for example
13. Mathematical rule
15. Greek philosopher famous for his paradoxes
19. Musical note between do and mi

Answer on page 310

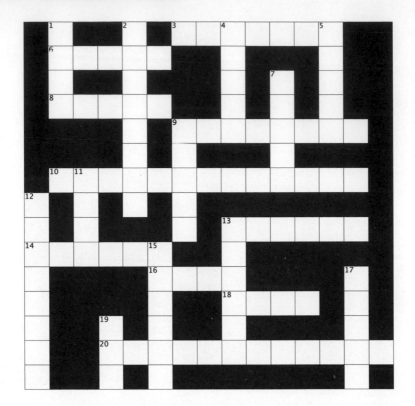

ACROSS

3. Meteorological condition
6. Solitary
8. Figure out
9. Fantasizing
10. Awareness
13. Substance
14. Disease caused by an uncontrolled division of abnormal cells
16. Finished
18. Proctor's call
19. What the "I" in "IQ" stands for

DOWN

1. The red planet
2. Everything there is
4. "As _____ so below"
5. Govern, as royalty
7. Person
9. Motivation
11. Portent
12. Germs
13. Finite
15. Droids
17. Get ready (for something)
19. Goal

Answer on page 310

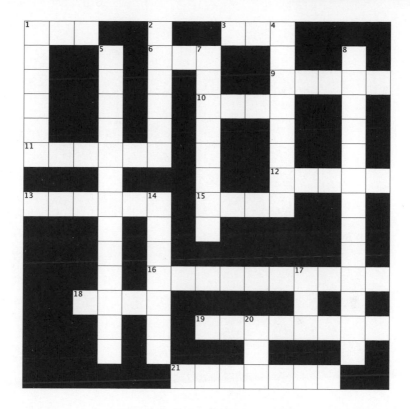

ACROSS

1. Collection
3. Word file type
6. Computer storage
9. Spotify offering
10. Like one end of a pool
11. State of inactivity
12. Ahead of time
13. Tables with facts and images
15. Opposite of far
16. Architectural diagrams
18. Alone
19. System of symbols
21. Illustration

DOWN

1. Pieces of paper
2. Diagrams like those used in coordinate geometry
4. Device often used to make graphics
5. Graphic designs
7. Exhibiting
8. Imagining
14. Signs such as "x", "@", "%" etc.
17. Bank fig.
20. Label

Answer on page 310

ACROSS
3. Lesson guideline
5. Subjects such as history and literature
6. Method used in chemistry class
8. Common adverb
11. Places where chemistry experiments are conducted
12. Damage
14. Discipline of quantity and space
17. Genre of creative writing, including novels, plays, poetry, etc.
21. Opposite of old
22. Follow
23. Courses

DOWN
1. Like sciences such as physics and chemistry, rather than anthropology or psychology
2. Study of the past
4. Music, dance, and painting, for example
7. Laptop or desktop
9. Musical theater
10. Funny performer
13. The science of plants
15. Chalkboard needs
16. Like a non-physical science
17. Opposite of high
18. Tests
19. Join
20. Written assignment

Answer on page 310

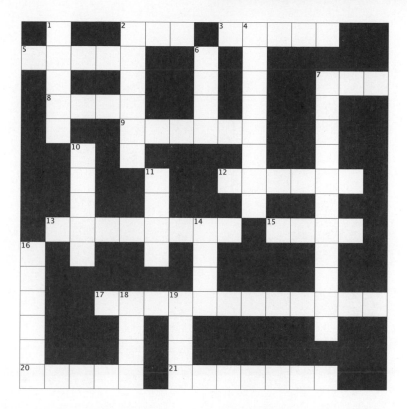

ACROSS
1. Creative activity
3. Connection in a series
5. H. G., who wrote *War of the Worlds*
7. At present
8. Standard
9. George, who wrote *1984*
12. William, who spread the word "cyberspace"
13. Artificially intelligent beings
15. Not fake
17. The monster created by Mary Shelley
20. Frontier that *Star Trek* aims to conquer
21. The "A" in USA

DOWN
1. Jules, who wrote *20,000 Leagues under the Sea*
2. Isaac, who wrote *I, Robot*
4. Robert, who wrote *Starship Troopers*
6. Famous sci-fi novel written by Frank Herbert in 1965
7. Name of Gibson's novel, which spread the word "cyberspace"
10. Groups
11. Norse god of thunder
14. Phillip K., whose book was the basis of the cult move *Blade Runner*
16. Humanoid machines
18. Athletic competition
19. Star showing a sudden large increase in intensity

Answer on page 311

ACROSS

2. Theorem integral to understanding triangles
8. Unit of electric power
9. Greek letter used for a famous statistical test
10. Computer memory
11. What the Sun generates
14. Ancient Greek mathematician
17. Terror
19. Hider's enemy
20. The final frontier

DOWN

1. The science of Newton or Einstein
2. Scientific explanations
3. Greek philosopher who wrote one of the first books on physics
4. Systematic arrangement
5. Albert, who came up with the theory of relativity
6. Isaac, who established the laws of motion and gravity before Einstein
7. The backbone of science
12. Luminous celestial bodies
13. Reasons
15. Profound
16. Unit of energy
18. Frozen water

Answer on page 311

ACROSS

1. Atomic particle without an electric charge
5. Darwin's "On the _____ of Species"
6. Atomic particle with a negative electric charge
7. Measure by which something changes, as in the _____ of radioactive decay
9. Opposite of negative
11. Appendage
12. Indefinite period of time
14. Atoms with the same number of protons and electrons but different physical properties; an example is "heavy water"
17. Unit consisting of atoms bonded together
19. A nuclear reaction that releases great energy
20. Emission of energy in the form of waves or particles; an example is X-rays
21. Grain in a J.D. Salinger title

DOWN

1. Central core of the atom
2. Chemical element, with symbol Ne, used in brightly glowing lights
3. Opposite of positive
4. Atomic particle with a positive electric charge
8. Particles which are fundamental constituents of matter, the spelling comes from James Joyce's *Finnegans Wake*
10. Atoms or molecules with a net electrical charge
12. A fundamental substance
13. Form of energy
14. Sick
15. Discharge from infected tissue
16. Atomic number for Hydrogen
18. Conclusion
28. Strong criticism (which you might give yourself if you forget common science terms)

Answer on page 311

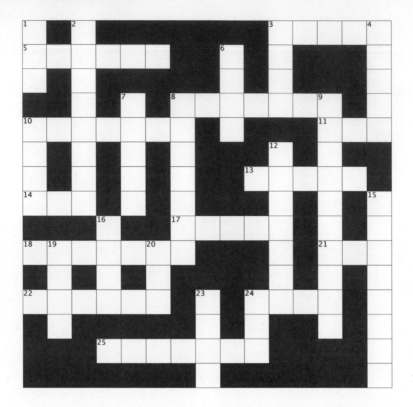

ACROSS

3. Spark
5. O_2
8. Purify a liquid
10. Radioactive form of an element
11. Commercials
13. Combination of two or more metallic elements, to give greater strength or resistance to corrosion
14. Star at the center of the solar system
17. Test _____, glass vials used commonly in labs
18. Dormant state
21. _____ the line
22. Fluid, such as water or milk
24. Shred
25. A flat, blunt instrument, used for mixing and spreading things, sometimes called a tongue depressor.

DOWN

1. Atom or molecule that has lost or gained of one or more electrons
2. H_2
3. Pacific retreat
4. Warms
6. Substance that turns litmus paper blue
7. State of matter, along with gas and liquid
8. Mass divided by Volume
9. Chemistry workplace
10. Part of the eye and the goddess of the rainbow
12. Seal
15. Periodic table listings
16. List of computer commands
19. Substance that turns litmus paper red
20. Jerk
23. Obscure
28. Coffee alternative

Answer on page 311

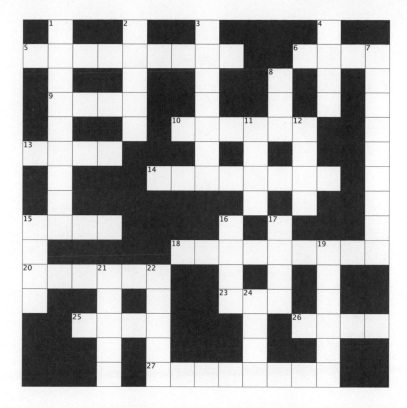

ACROSS

5. Etna and Vesuvius, for example

6. Dark rock used mainly as a fuel

9. Toy flown in the wind by holding a string attached to it

10. Prehistorical remains, such as bones, wood, or organisms

13. Something used in solving a crime or mystery

14. High pH

15. Pointed top of a mountain

18. Rapidly falling mass of snow down a mountain, getting bigger as it comes down

20. Heavy object used to moor a vessel to sea bottom

23. Realize

25. The inner center of the Earth

26. Substance found in deserts and on beaches

27. Matter on the surface of the land that may in time become consolidated into rock

DOWN

1. Danger on a mountain

2. Caverns

3. Layer of Earth lying under softer material within the Earth's crust

4. Precious metal

7. A common carbonate sedimentary rock

8. 22nd letter of the Greek alphabet

11. Common name for sodium chloride

12. Highway division

15. Design

16. The "m" in $E=mc^2$

17. Speed

19. Metal found in meteors

21. Fishing needs

22. Coral ridges

24. Test

Answer on page 311

ACROSS

1. Hard, often shiny, materials including lead, and gold, for example
4. The main substance of tree trunks or branches
6. Made of clay that has been hardened by heat
9. Round solid figure
11. The space above the Earth, as in "the night _____"
12. A hard transparent substance, typically breakable
13. Massachusetts' Cape _____
15. Brawn or power
17. Synthetic organic materials used as plastics and resins
20. Like metals shaped by hammering, as in "_____ iron"
22. Give off
24. Matter, substances, things, as in "building _____"

DOWN

1. Natural inorganic substances, such as iron, manganese, or copper, for example
2. Large brass instrument
3. "Pine _____," rounded fruit of a pine tree
5. Of little thickness
7. A large amount of matter
8. A mineral, such as quartz
9. Geometric form, such as a square or triangle
10. Pest
14. Mass divided by volume
16. Piece of farm equipment
18. Univ. in Cambridge
19. Alloy of iron; often used in high-rise buildings
21. Fetch, acquire
23. Crater

Answer on page 311

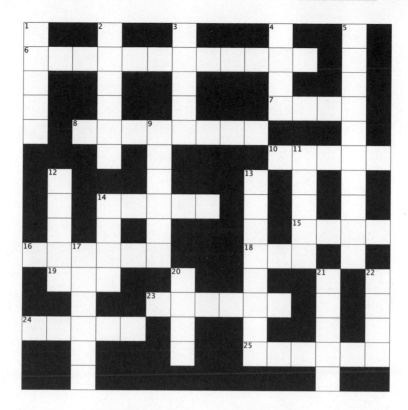

ACROSS

6. Increase in rate of speed
7. Back
8. The impetus of a moving object. Newton's second law of motion states that the "time rate of change of _____ is equal to the force acting on the particle"
10. The final frontier
14. Effect on the motion of a body; might or pressure
15. Celestial body visible on a clear night
16. Hypothesis
18. Feminine pronoun
19. Glimpsed
23. Attractive metal
24. Put on a scale
25. Standard

DOWN

1. Our planet
2. Sir Isaac, who studied gravity experimentally and mathematically
3. Locomotive that is used to illustrate gravity and relativity principles by analogy
4. The square root of sixteen
5. Gravity is a force of _____ between any two bodies
9. The "E" in $E = mc^2$
11. Whole number that is divisible only by 1 and itself
12. Sounds of resignation
13. Albert, who authored the theory of relativity
14. The movement of water or electricity, as in a current or stream
17. With little effort
20. Plummet
21. Shapes
22. Actual

Answer on page 312

ACROSS

2. Transparent

4. Chemical element Ne, used in fluorescent lighting

6. Chemical element H_2, used in petroleum refining and fertilizer production

9. Li, a soft silver-white metal

10. Unknown author, for short

11. A natural valuable solid material, as in "iron ____,"

12. Center, nucleus

14. An ancient prophet, seer

16. Refine

17. Chemical element with the symbol "He;" used as a gas for lifting balloons and airships

18. Chemical element Si, used in computer chips and solar cells

19. Chemical element Ca, required for bone health

20. Atmosphere, sky

22. Chemical element N, used as a coolant

24. Chemical element Fe, used to make steel

25. Chemical element S, used in making gunpowder

DOWN

1. Chemical element O_2, required for breathing

2. Chemical element C, its compounds form the basis of all living things

3. Chemical element Na; when combined with chlorine it produces salt

5. Chemical element F, in a compound it is used to maintain dental health

7. Chemical element AU, considered the most precious of metals

8. Chemical element N, used to make fertilizers and dyes

10. Basic unit of a chemical element

13. Unit of electrical resistance

15. Chemical element Cl, occurring mainly as sodium chloride in seawater and salt deposits

17. Chemical element Ho, used in nuclear reactors

18. Salt water body

19. Automobile

21. Chemical element Sn, used commonly to make cans and containers

23. Hairstyling product

Answer on page 312

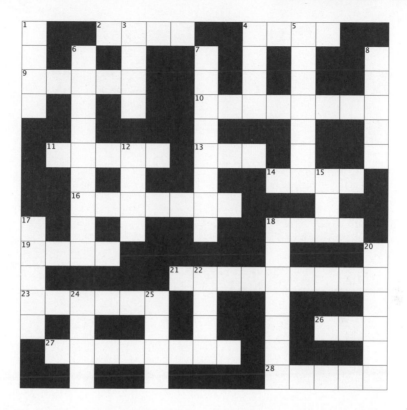

ACROSS

2. Quantity chosen as the norm or standard
4. Identical
9. Basic units of matter
10. Subatomic particle with a negative charge
11. Attractions between atoms or molecules that form chemical compounds
13. Atom with either a negative or positive charge
14. Actual
16. Type of energy
18. Greasy
19. Quantity of matter
21. Like a mineral
23. Tools used in drilling and eye surgery
26. Animal park
27. Chemical bonds of electrons between atoms
28. Animal with humps, which can survive for long periods without food or drink

DOWN

1. To guide, show the way
3. Org. behind Perseverance
4. Secure
5. A material made of two or more substances
6. Combinations of two or more elements
7. Any substance consisting of matter
8. Skeleton parts
14. Ventilation passageway
15. Everything
17. Little
18. Chemistry branch
20. Group of fish
22. Gas used in some bright lights
24. "On the ____;" immediately
25. Common name for sodium chloride

Answer on page 312

ACROSS

4. Groups of atoms bonded together
6. It turns red litmus paper blue
8. Combinations of two or more elements
10. Ill
11. They can have a positive or negative charge
12. Paper for determining pH
15. Remain
16. Of the mind
17. Opposite of dirty
19. Substance that increases the rate of chemical reactions
23. Blasts, as caused by bombs
25. Rearrangements of the molecular structure of a substance

DOWN

1. Substances in the periodic table
2. Device consisting of a weight suspended from a pivot so that it can swing freely back and forth
3. Cylindrical glass container for laboratory use
5. Leader of a corp.
7. They turn light blue litmus paper red
9. Combine
13. Copper, gold, or iron, for example
14. Kind of energy or reactor
18. Level
20. Concerning
21. Opposite of no
22. Test _____, long U-shaped glass or plastic containers used commonly in laboratories
24. Unwanted email message

Answer on page 312

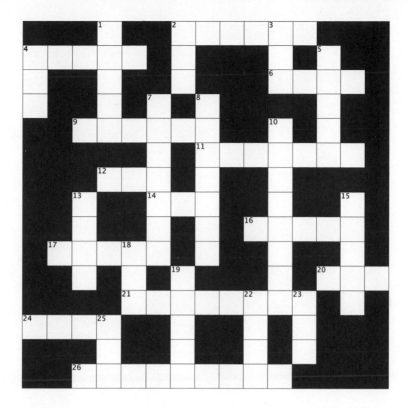

2. One of the three main states of matter in addition to gasses and solids
4. The most common noble gas, with chemical symbol Ar
6. Orderly
9. An inert gas, with chemical symbol He, which is the lightest of the noble gasses
14. Symbol for the chemical element fermium
11. The main constituent of natural gas; its chemical formula is CH_4
12. Sticky organic compound secreted by honeybees
14. Vacation spot
16. Chlorine is partially this color
17. Gas that makes up a protective layer of the earth's stratosphere
20. Valuable rock
21. Unreactive, colorless, and odorless gas that makes up about 78 percent of the Earth's atmosphere
24. Gasoline, for example
26. Easily set on fire

1. A gas such as neon, helium, or argon; their name can also mean aristocratic
2. Allow
3. An atom or molecule with either a positive or negative charge
4. We breathe it all the time
5. A noble gas, with chemical symbol Rn
7. O_2; as in "carbon ____"
8. A colorless gas with a pungent odor, used commonly in fertilizers; chemical formula: NH_3
10. Its chemical symbol is Cl; it is used, for example, in swimming pools as a disinfectant
13. Gush or pour forth
15. Chemically inactive
18. Oui's opposite
19. Water vapor, condensation
22. Aim
23. Web
25. Texter's reply

Answer on page 312

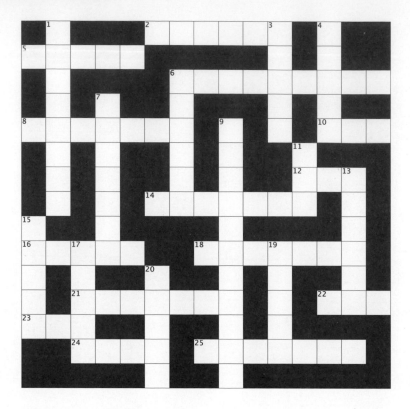

ACROSS

2. Liquids
5. H_2O
6. The measure of a fluid's resistance to flow
8. Intoxicant
10. Organ with an anvil
12. We breathe it in all the time
14. Chemical element with symbol Hg; used commonly in older thermometers
16. Tool with a beam
18. Red toxic liquid with an irritating smell, occurring chiefly as salt in seawater and brines; its chemical symbol is Br
21. A colorless volatile flammable liquid
22. Volcanic residue
23. Body of water such as the Mediterranean, or Caspian
24. Build
25. Creeks or rivulets

DOWN

1. Liquid used as fuel for internal combustion engines
3. Mar or disfigure
4. Strong solution of water and salt
6. Space enclosed by a closed surface
7. Able to be dissolved, especially in water
9. Liquid that contains ions, found in some sports drinks
11. Inlet of the sea, as in the "_____ of Biscay"
13. Natural streams of water flowing to the sea, a lake, or ocean
15. Runs, as water
17. Water vapor
19. Perhaps
20. Bodies of water surrounded by land

Answer on page 312

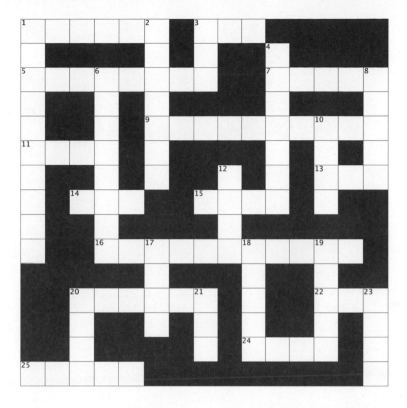

ACROSS

1. The action of moving
3. Unit of electrical resistance; symbol: W
5. Any substance consisting of matter
7. Warms
9. Charged
11. Gas used in many a sign
13. It contains a mixture of oxygen and nitrogen
14. A/C measure
15. Gas, coal or other substance used to produce energy
16. Able to catch fire easily
20. Units of energy
22. Car tank filler
24. Boggy, acidic ground used as fuel; also used in gardening
25. Marine formation

DOWN

1. Automatic
2. Type of energy generated in a reactor
3. Petroleum liquid
4. Relating to heat
6. Attractive
8. Renewable kind of energy
10. Dark rock that is mined as a fuel
12. Operates
17. Odometer unit
18. Garden bulb
19. Ignite
20. Taunt
21. What planets orbit
23. Luminous body in the sky

Answer on page 313

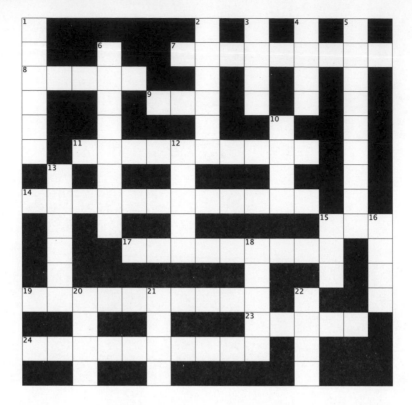

ACROSS

7. Connection speculated to exist between widely separated regions of space

8. Moisture, water vapor

9. What eyes allow us to do

11. Like planets or moons beyond our sun's pull

14. Situated between stars

15. Opposite of new

17. Darwin's theory

19. They travel to outer space

23. Saturn's largest moon

24. Instruments that allow us to see distant objects in space

DOWN

1. The universe (seen as a whole)

2. Celestial objects with a tail of gas and dust particles, the most famous of which is named after astronomer Edmund Halley

3. Discharge

4. The Point on the surface of the Earth where the lines of longitude meet, once in the north and once in the south

5. Of the sky

6. Systems of stars, like the Milky Way and Andromeda

10. American agency responsible for the space program: abbr.

12. Prefix meaning "the stars"

13. In geometry, these are measure in degrees or radians

15. The first integer

16. Word before space or end

18. Tries

20. Say, relate

21. Pleasant

22. An incandescent celestial body, such as the sun

Answer on page 313

ACROSS

1. A liquid compound
4. Smell
7. Liquefied by heat
9. Substance made by combining other substances
11. Amount of a given substance in a solution; focusing one's attention on something
13. Victory
14. Cheese in a Greek salad
15. Hot drink, often offered as an alternative to coffee
17. Action word
21. Also called ethanol; an intoxicating constituent of beer and wine
23. Sticky
24. Soaked, permeated
25. Dairy product that is normally pasteurized

DOWN

1. Common name for sodium chloride (NaCl)
2. The least number of atoms in a molecule
3. Opposite of in
5. Liquefied
6. Solution of acetic acid; used as a condiment
7. Softening
8. Neither acid nor alkaline
10. Hot drink, often offered as an alternative to tea
12. In a natural state
16. Drinks obtained from fruits
18. Set on fire
19. River rental
20. Beer or coffee topper
22. Fat

Answer on page 313

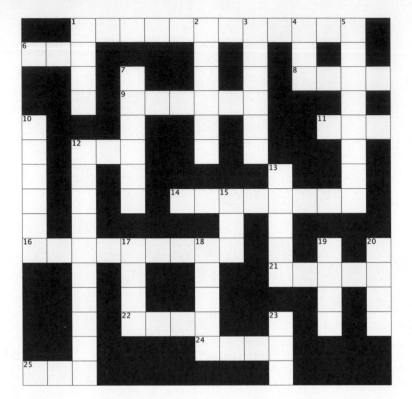

ACROSS

1. "_____acid," acidic solution with the chemical formula: HCl
6. A common round baked good
8. Opposite of light
9. Favorite food of Bugs Bunny
11. It can be of positive or negative charge
12. A negative adverb
14. Acid made by oxidizing sulfur dioxide; used in industry, mainly in fertilizers, and laboratory research
16. Caustic
21. Float
22. Actual
24. Secure
25. Insect that makes honey

DOWN

1. Thermodynamic energy
2. Acid present in lemons
3. Related to milk
4. One of the two colors that litmus paper turns to
5. Acid formed when carbon dioxide dissolves in water; chemical formula: H_2CO_3
7. Acid that gives vinegar its taste
10. Corrosive yellow acid made with nitrates and sulfuric acid; chemical formula: HNO_3
12. Deactivate
13. Clear
15. Falsehood
17. Opposite of under
18. Containers often found in a lab
19. Common dairy product
20. Truncated
23. Subject of an 18th century "party" in Boston

Answer on page 313

ACROSS

1. Infrequent, exceptional
2. Seasons
6. Forces of attraction holding atoms together in a molecule
8. Weather conditions characterizing a region
9. Kind of chemical bond
11. Smudge
12. Second word in many a fairytale
11. The Carrier of genetic information
15. H_2O
17. Try again
19. "_____ acid," a carboxylic acid that can be saturated or unsaturated
20. Everything
21. What doesn't gather on a rolling stone
23. Lived
24. Cylindrical metal container
25. Upcoming, following
26. Referred

DOWN

1. Chemical process whereby one substance changes into other substances
2. Symbol for the chemical element radium
3. Common name for Na_2CO_3
4. Combination
5. The fundamental units of chemical compounds; groups of atoms bonded together
7. Opposite of short
10. Like minerals
13. Complex molecules that are essential to building muscle mass and other parts of the body
14. Chemical bonds formed by electrons
16. Metals made by mixing other metallic elements
18. A chemical used commonly to sterilize materials, or to whiten them
22. Identical

Answer on page 313

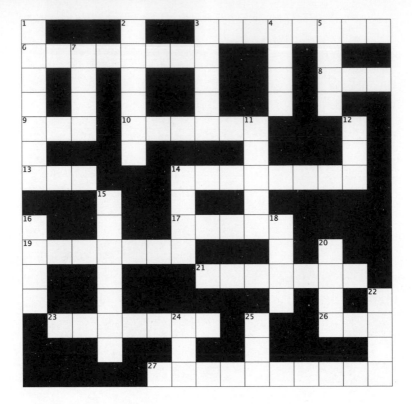

ACROSS

3. Transporting goods or materials

6. The ocean between Europe and Africa and North and South America

8. "Lake ____," in Tasmania; open area of grassy land

9. Short-lasting trend

10. Ocean surrounding the North Pole

13. Taxi

14. Amount of salt dissolved in a body of water

17. Aquatic organisms capable of photosynthesis

19. Attackers of ships

22. Activity at sea

24. Ocean plant

27. Number of binary digits

28. Captain's concern

DOWN

1. The world's largest ocean

2. The youngest of Earth's oceans

3. Sir Robert ____ (1868-1912), English explorer and naval officer who made a journey to the South Pole by sled

4. Brown material consisting of decomposed vegetable matter that's often used in gardening

5. A small island

7. Opposite of sea

11. Sea that is part of the Pacific Ocean in Asia

12. Inlet

14. "The high ____;" the open ocean

15. Sea in the northwestern part of the Indian Ocean

16. Opposite of close

18. Snake-like fish

20. Effortlessness

21. Unit of speed equal to one nautical mile per hour

23. Celestial body that comes out at night

25. Age

26. Fall behind

Answer on page 313

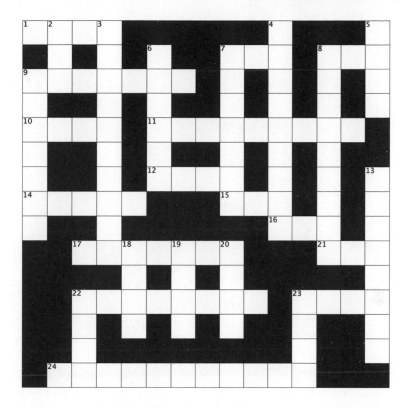

ACROSS

1. Raw facts
7. Prefix meaning three
8. Opposite of cold
9. Weather or atmospheric conditions that characterize a region
10. Opposite of more
11. Wet weather event
12. Type of wind blowing from one direction
14. Opposite of over
15. Ski slope
16. _____ Rummy
17. Meteorologist's concern
21. The center of a storm
22. Severe snowstorm with high winds
23. Greek Earth goddess
24. Measurement taken with a thermometer

DOWN

2. Everything
3. It surrounds the Earth, containing gases such as oxygen
4. Electrical discharge between a cloud and the ground, during a storm
5. The opposite of start
6. Supplies
7. Sound heard during some weather events
8. Violent storm, in particular in the Caribbean
9. Unit of temperature
13. Weather prediction
18. Very dry
19. Mist, fog, cloud
20. Exceptional, unusual
22. A bundle of hay or cotton
23. A very strong wind

Answer on page 314

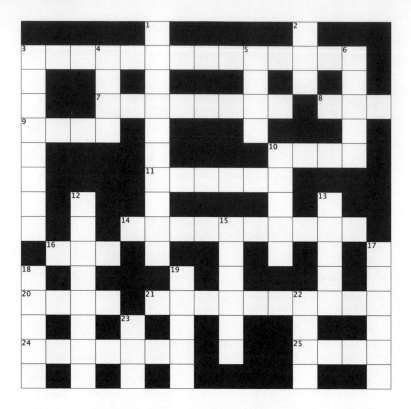

ACROSS

3. Branch of physics that deals with heat
7. Device that emits heat
8. Opposite of dry
9. Personal identifier
10. Opposite of cool
11. Artic plain
14. Term describing the transfer of heat
16. Cunning
20. Consistent speed or rate
21. Scale of temperature on which water freezes at 32°
24. Scale of temperature on which water freezes at 0°
25. Material such as gas, coal, or oil, burned to produce heat

DOWN

1. Transmission of heat through a substance
2. Place for outdoor cooking
3. Conduction of heat
4. Exceptional
5. The basic unit of all matter
6. Water vapor
10. Refuse
12. A group of atoms bonded together forming the basic unit of a compound
13. Unit of work or energy
15. Vitality
17. Opposite of big
18. Open area
19. Opposite of lower
22. Weight
23. Symbol for the chemical element nickel

Answer on page 314

ACROSS
6. Broadcast
8. Consonant sound
9. Sound producers
11. Witticism
13. Quality of sound
15. Type of fork that emits a musical note
16. Require
17. Orbital period
19. A collection
20. Place for a mud bath
24. Vibrant, like a voice
26. Thunderous
28. Piece of equipment for a rock band
29. Listen to
30. The art of Beethoven and Beyoncé

DOWN
1. Oscillations of sound waves
2. Unpleasant sound
3. Dull, heavy sound; thump
4. Rate of a wave's vibration
5. Musical sound of a certain pitch
6. The science of sounds
7. Climb
10. Resonances
12. Phone purchase
14. Unit of frequency; symbol: Hz
18. Reflected sound
21. High playing card
22. Exceptional
23. Lenient
24. Poles
25. Rainproof cover
27. Darken

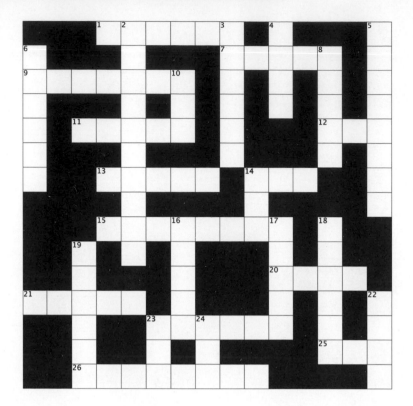

ACROSS

1. Devices that generate intense beams
7. Device used to manually turn an axle
9. Things
11. A point where parts of machines come together
12. Charge
13. Toothed wheels that alter the speed of a driving mechanism or vehicle
14. Engine part
15. "Ball _____," part of a machine
that counteracts friction between rotating parts and their housing
20. Rod or shaft for a rotating wheel
21. A rigid bar on a pivot, hinge, or fulcrum, used to help move heavy loads
23. Lend a hand
25. Although
26. Thrilling

DOWN

2. Greek mathematician and inventor,
known broadly for discovering a principle while taking a bath, shouting "Eureka," according to legend, after his discovery. He designed innovative machines, such as his screw pump
3. Threaded hardware
4. Device for lifting objects, used commonly
to change an automobile tire
5. A bar that slides into a socket in a door or window
6. Movement
8. Cutting utensil
10. Collection
14. The tooth of a gear
16. Lassos
17. Elevator enclosure
18. Weight lifting device
19. Contraption
22. Car
23. Circle section
24. Title for Isaac Newton

Answer on page 314

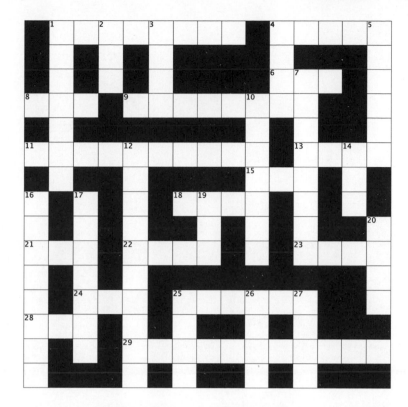

ACROSS

1. Location, place, locus
4. It changes the motion of a body
6. A lyric poem
8. Purpose, function
9. Speed
11. Speed up
13. Apart, at a distance
15. Engine stat.
18. Opposite of slow
21. Knight's title
22. Change
23. A long legendary tale
24. Works in a gallery
25. Sir Isaac, who is known for his laws of motion
28. It usually accompanies "neither"
29. Movement back and forth, like the movement of a pendulum

DOWN

1. Science that deals with motion
2. Large body of water
3. Variable in determining the distance something has gone
4. On _____, walking
5. Vitality, strength
7. Branch of mechanics dealing with motion
10. State of rest or uniform motion
12. Movement from place to place
14. Everything
16. It is calculated by multiplying the rate of movement by the time taken (D = R ´ T)
17. Opposite of backward
20. The "R" in the formula D = R ´ T
25. Pleasing
26. Story
27. Musical tone

Answer on page 314

ACROSS

1. Branch of physics that deals with transistors, and microchips
6. It can be positive or negative; the coulomb is the unit used to measure it
7. It is normally preceded by "neither"
8. A "_____ circuit" occurs when there is too much electricity in a circuit
9. Power
12. Opposite of enter
14. Select
16. A switch that opens or closes a circuit
17. Term used to refer to the energy that surrounds electrically charged particles
19. Additional
20. Stitch
21. Opposite of murky
22. Subatomic particles with a negative charge

DOWN

2. The path in which electric current flows
3. Web
4. Fly high
5. Electrical safeguard
6. Stream
7. Observe
10. Current units
12. Opposite of exit
13. Units of electrical resistance
15. Number expressing the central value in a set of data
17. Payment for a professional service
18. Electrical discharge igniting an internal combustion engine
20. Canine command
21. Trig function

Answer on page 314

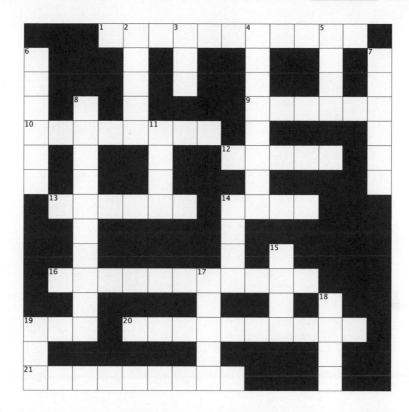

ACROSS

1. Wavelength present in sunlight
9. Elevates
10. Band of colors, as in a rainbow
12. Triangular transparent object that separates white light into colors
13. Devices used to repair retinal detachments
14. Smell
16. The process by which light waves are spread out after passing through a narrow aperture
19. The number base for common logarithms
20. Image seen in a mirror
21. Health care profession of examining the and prescribing corrective lenses

DOWN

2. Opposite of dark
3. It emanates from the sun, or other luminous body
4. A wavelength greater than that of the red end of the spectrum
5. Watches
6. Corrective _____, what the optometrist prescribes
7. The faculty of sight
8. Deflection of light as it passes obliquely through some medium
11. Opposite of front
14. Leave out
15. Publish on a social media site
17. Shade
18. Central points
19. Number of binary digits

Answer on page 315

ACROSS

6. Substance that provides thrust in a rocket engine
9. Buddhist state
10. Front page information
11. Small cylindrical piece that is used to hold things together or hanging things
14. NASA spaceflight program named after a Greek god that landed the first humans on the moon
15. Vehicle controlled by remote control that is used to travel over extraterrestrial terrain, such as on the moon or Mars
17. Confidence
19. Opposite of over
23. Rocket engine
24. Data
25. Construct

DOWN

1. Any vehicle for travel beyond Earth
2. Spaceflight program initiated by the Soviet Union and continued by the Russian Federation
3. US spaceflight program with human crews that shares a name with a zodiac sign
4. Power source
5. Technology and industry for astronauts
7. The rocket's "rise" from a launch pad
8. Zenith
12. A specific location or position
13. What we eat
16. Move backward
18. Powerful kind of engine
20. Opposite of west
21. Vessel
22. "_____ me up, Scotty!"

Answer on page 315

ACROSS

5. Device that measures temperature
7. Appeal for assistance (exclamation)
9. Having a high temperature
10 Paddle
11. Temperature scale on which water freezes at 32°
15. Blows on
17. Unit of temperature measurement
18. Relating to low temperatures
20. Length of time;
23. Cloudless
24. Temperate
25. Movement
26. Hue, color variety

DOWN

1. The temperature (Celsius) at which water freezes
2. Area with a distinct climate
3. Warmth
4. Unit of temperature in the International System of Units; symbol: K
6. Temperature scale on which water freezes at 0°
8. Place or situate
9. Catch
12. Heat transfer by the flow (velocity) of a fluid; often compared to *convection*
13. Opposite of urban
14. Argument or evidence that is used to establish something
16. "_____ zero," temperature at which all atoms completely stop; it corresponds to the number zero on the Kelvin temperature scale
17. Period when the sun is out
19. Shadowy
21. Period when the sun is not out
22. A dark are on the sun's surface

Answer on page 315

ACROSS

6. Model referring to the sun as the center, developed substantively by Copernicus
7. Informal
10. Copernicus, for one
12. Beam of sunshine
13. Copernicus is the _____ name of Mikołaj Kopernik, a common linguistic practice in his era
15. Collection of objects
17. Wrinkle
18. Sizzling
20. Planet seen as the center of the cosmos in geocentric theory
24. Star at the center of the solar system
25. Johannes _____, who discovered three laws of orbital movement
26. Ancient Greek astronomer whose geocentric theory remained dominant until Copernicus

DOWN

1. The theory that Copernicus' theory replaced
2. Unprecedented
3. Fine, good
4. Semicircle
5. Copernicus was also interested in this branch of knowledge, concerned with wealth
8. Copernicus' nationality
9. The curved path of a celestial body, such as the Earth around the sun
11. Roman sun god; same word is used in Spanish for the actual sun
14. Italian Renaissance astronomer and physicist, who discovered evidence to support Copernicus' theory
16. Give off
19. Opposite of closed
21. Second chance
22. Sacrosanct
23. Firmament
24. Drain

Answer on page 315

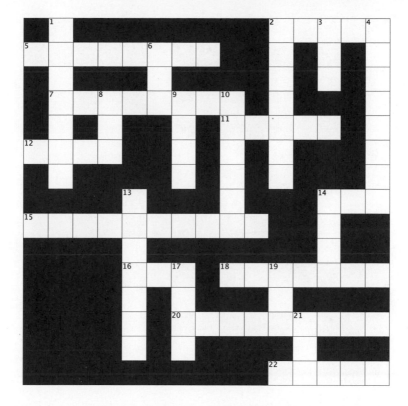

ACROSS

2. Electricity
5. Literally, a "small neutron," in Italian, the term given by Fermi to a neutral subatomic particle with a mass close to zero hypothesized earlier by Wolfgang Pauli
7. In 1939, Fermi became professor of physics at this New York City university
11. Topic or subject
12. Connect
14. Make a break for it
15. The science of data and the inferences that can be gleaned from it
16. International code of distress
18. Branch of physics for which Fermi became well known; relating to the nucleus of the atom
20. The working parts of something
22. Warn

DOWN

1. Fermi created the first nuclear _____, where nuclear chain reactions can take place
2. Fermi was a leader on the Manhattan _____, which helped develop the atomic bomb
3. The bomb brought an end to World _____ II
4. Fermi conducted the first nuclear chain _____
6. Computer co.
8. Ran
9. Weapon Fermi helped develop
10. Infinitesimal
13. Process in which the nucleus of an atom splits
14. Fermi was born in this capital city of Italy
17. Identical
19. Covert org.
21. Nothing, void

Answer on page 315

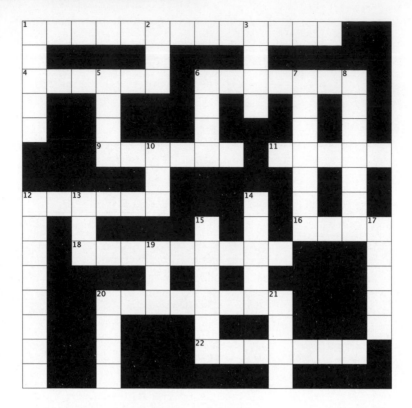

ACROSS

1. Spontaneous emission of radiation in the form of particles or high energy photons resulting from a nuclear reaction

4. Type of number

6. Type of nuclear reaction, contrasted with fusion

9. "Nuclear _____," an alternative form to oil, gas, wind, or solar

11. Numerical data

12. Type of nuclear reaction, contrasted with fission

16. Large divisions of time

18. Heavy hydrogen

20. Subatomic particle with a negative charge

22. Central region or core of the atom or of the cell

DOWN

1. Undergo a chemical or physical change

2. Curve, semicircle

3. Huge

5. Create

6. Shape something by heating it

7. Elemental form

8. Atomic particle without a charge

10. Epoch

12. Ease or smoothness

13. Upset

14. Dwarf planet passed Neptune

15. Subatomic particle with a positive charge

17. Small flash of light discharged by electrical discharge

19. Golf accessory

20. Simplicity

21. Nothing, nil

Answer on page 315

ACROSS

2. Opposite of order
7. Earth, Mars, and Mercury, for example
9. Something the requires a solution
10. Type of stadium
12. "____ firma," dry land
13. Recharge
17. The science or study of the origins of the universe
18. State of rest or uniform motion
20. Greek philosopher and mathematician who coined the phrase "music of the spheres;"
22. Immensity

DOWN

1. Star systems
2. Polish astronomer who institutionalized heliocentric theory
3. Basic particles of matter
4. Of the sun
5. Cosmos
6. Discharges, gives off
8. Celestial bodies
11. Soviet space station, launched in 1986and terminated in 2001
14. Carl ____ (1934–1996), American astronomer, who wrote the novel *Contact* (1985) and was in the TV series, *Cosmos*
15. Having only mass or quantity, not direction
16. Sagan's novel turned into a movie in 1997
19. Exploding star
21. Upper part

Answer on page 316

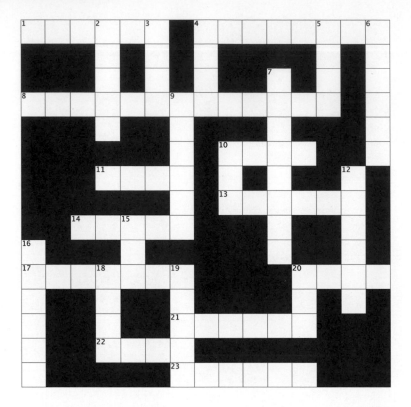

ACROSS

1. Superheated matter, as on the Sun
4. Gives off, as heat
8. Emission of light form a hot body, like the Sun
10. Where nuclear reactions occur in the Sun
11. The Sun is one of these celestial bodies
13. The glow around the sun; from Latin "crown"
14. The _____ Way, the galaxy in which the Sun is located
17. "Solar _____," whereby the Sun is obscured by the moon
20. Earth's natural satellite
21. A radioactive metal; symbol Ra
22. What the Sun generates
23. Inert gas, produced in stars and the second most abundant element in the universe after hydrogen

DOWN

2. Of the Sun
3. Dry, scorched
4. The Sun's emissions
5. Woody plant with leaves and branches
6. "Solar _____," the planets and their moons in orbit around the Sun
7. Having a wavelength greater than that of the red end of the spectrum; often contrasted with "ultraviolet"
9. The "E" in $E = mc^2$
12. Chemical element with symbol C,
15. Circuit
16. The Sun god in Greek mythology
18. 1/12 of a foot
19. Third planet from the Sun
20. Quiet

Answer on page 316

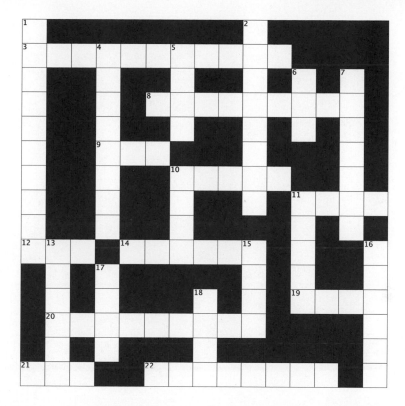

ACROSS

3. It can be measured in Celsius or Fahrenheit degrees
8. Regions of the Earth occupied by living organisms
9. The body at the center of the solar system
10. Kind of bear
11. Mound of sand formed by the wind
12. Consume
14. Purpose
19. Early morning
20. Moderate (climate)
21. Frozen water
22. Mugginess

DOWN

1. Layers of gases surrounding a planet
2. Kind of storm
4. Word after air or atmospheric
5. Very dry
6. Opposite of dry
7. Summer, fall, winter, and spring
11. Stunned
13. Area around the North Pole
15. Domesticated
16. Frozen land area
17. Period, as in "daylight saving ____"
18. Opposite of cold

Answer on page 316

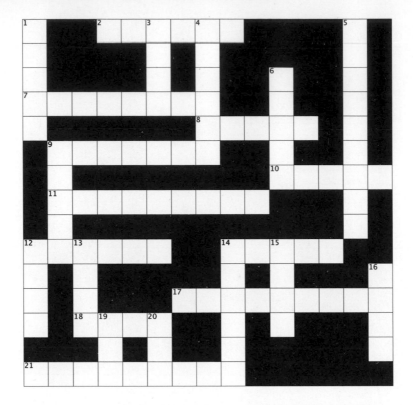

ACROSS

2. The science of sight and light, studied by Aristotle

7. The basic things of matter, also the name of Euclid's book of mathematics

8. Our planet, which Aristotle described as round

9. The science of matter, discussed by Aristotle in a book with this title

10. Lines from the center to the circumference of a circle

11. The science of celestial bodies and space, also studied by Aristotle

12. Movement

14. Reasoning according to rules, elaborated by Aristotle

17. Aristotle was born on the border of this state, of the Balkan Peninsula

18. Make

21. Aristotle's term for this type of logical format: "All humans are mortal; I am human; therefore I am mortal"

DOWN

1. Aristotle's nationality

3. Opportunity

4. Reasons

5. Logical conclusion

6. One of the four elements posited by Aristotle

9. Aristotle's teacher

12. Fable

13. Proctor's announcement

14. School that Aristotle established in Athens

15. Divinities

16. Foremost

19. Everything

20. Bother

Answer on page 316

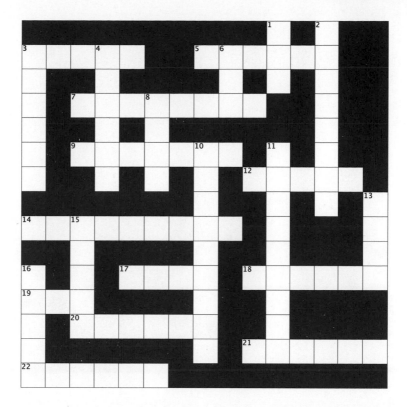

ACROSS
3. Mist, steam
5. Water, physically
7. The "H" in H_2O
9. Compactness; of water it is 997 kg/m³
12. Ice crystals on the ground
14. Thickness of a liquid, such as water
17. Coal, petroleum, or gasoline, for example
18. Add water to a liquid, so as to make it weaker
19. Colonnade tree
20. They need to be watered to survive
21. Facet, component
22. Annually

DOWN
2. The Earth orbits it
2. Unscented
3. Capacity
4. The "O" in H_2O
6. Frozen water
8. Sign of corrosion
10. Offensive
11. Living things
13. Vivacity
15. Overwhelm
16. Rot

Answer on page 316

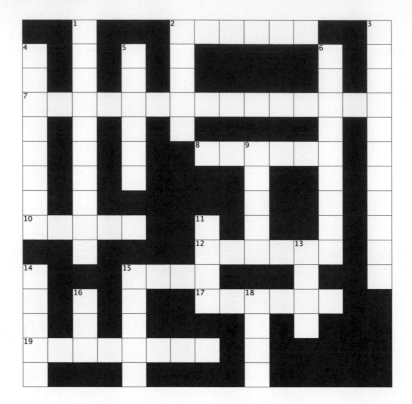

ACROSS

2. Wax light source
7. Branch of physics that deal with the force between charged particles
8. Sir Isaac, who studied light and its properties scientifically
10. Ancient tales
12. Light consists of _____ quanta called photons
15. Epic tale
17. It shows your image
19. Albert, who connected light with mass and energy in his famous equation $E = mc^2$

DOWN

1. The distance between consecutive corresponding points of the same phase
2. Red or green, for example
3. Charged
4. Visible _____, contains the colors that the human eye can see
5. Layer at the back of the eye that is sensitive to light, that can become detached
6. Concentration
9. The color of snow
11. Ray
13. Radiance, shine
14. The color of grass
15. Dots that appear occasionally on the surface of the Sun
16. Star at the center of the solar system
18. Percentage, ratio

Answer on page 316

ACROSS

1. Geometric solids that refract light into colors
6. The color of grass
7. A color between red and blue; that symbolized rank or privilege in Ancient Rome
9. Full range, as colors
12. Rosy color
15. Opposite of black
16. Like blues, greens, or purples
19. Color of wood or soil
20. Brewed drink
21. Lackluster
22. Color between green and orange that's used to describe sensationalistic journalism
23. Green stone used for jewelry

DOWN

2. Region of the spectrum greater than red; also, contrasted with "ultraviolet"
3. Referring to color sensations
4. Color or shade
5. The color of blood
7. Coloring matter
8. Warm pink color; also, a common flower given at Valentine's
10. Shade
11. Spring flower
14. Bit of color
15. Make a mistake
17. Fruit named for its color
18. &
19. Color of the sky

Answer on page 317

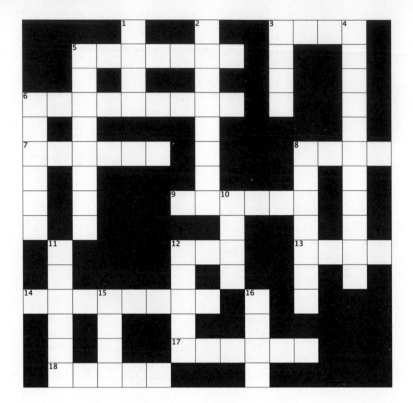

ACROSS

3. Opposite of under
5. Like a volcano
6. Jagged
7. A four-sided figure with all sides equal
8. Half a quart
9. Two-dimensional
12. Chart type
13. Average
14. Curved sickle shape of the moon
17. The shape of the Milky Way
18. Circumference

DOWN

1. Tight-fitting
2. Like straight lines that never cross
3. Egg-shaped
4. Four-sided figure
5. Ring-shaped
6. Exclusive
8. Three-dimensional triangle
10. Peak, summit
11. Cello feature
12. Sheets of glass in a window
15. Luminous celestial body
16. Bicycle part

Answer on page 317

ACROSS
1. Round
7. Opposite of interior
8. One of two children at the same birth
9. Part of a bridal outfit
10. Smooth, horizontal
12. It is 180°
14. Smooth, level
15. Feel of a surface
17. Swallow
19. Opposite of rough
21. Frontage
23. The shortest possible line between two points on a sphere or other curved surface

DOWN
2. Pocked
3. Arc
4. Overlay with some protective substance or material
5. Creative activity such as painting
11. Precise
12. Coating; covering
13. Distant
15. Six-sided figure
16. Through, by means of
18. Communal
19. Part of a square
20. Miniscule
22. Cupola, like the type used in some churches

Answer on page 317

ACROSS

1. Shooting stars
6. Frozen water
7. Opposite of big
11. Winter fight ammunition
12. Ogle
15. Comets have this "appendage,"
17. Celestial body
18. Metal found in meteors
19. Appearing at intervals
21. Infrequent
23. Worth
24. Go around, which some comets do with the Sun

DOWN

2. _____ after
3. Edmond _____ (1656–1742), English astronomer known for identifying a bright comet named after him
4. Period, era
5. Otherworldly
8. Red planet
10. Bring up the rear
13. Opposite of west
14. Type of dirt of which comets are made
16. Belt in space
19. Of the sun
21. Rock
22. Disturbance

Answer on page 317

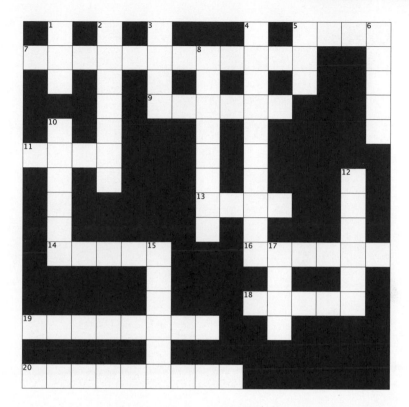

ACROSS

5. Nights before holidays
7. Luminous, emitting light (like meteors)
9. "Meteor _____," referring to a number of meteors radiating from one point in the sky
11. Hard silver metal, with chemical symbol Fe, found in meteorites
13. Memo
14. Stones, which is what meteors are
16. What a meteor looks like
18. Celestial body
19. Like some shiny material
20. The science of outer space and celestial bodies, including meteors

DOWN

1. The first number
2. Coming down to Earth
3. Date in March to beware
4. Debris from space that survive passage through the Earth's atmosphere, striking the ground
5. Greek H
6. Constellation elements
8. "_____ star," a meteor burning up as it enters Earth's atmosphere
10. Large cavity that meteors leave in the ground
12. Metal found in meteors
15. Taken
17. Useful item

Answer on page 317

ACROSS

1. The moon orbits this planet
7. Follow
8. Stages
9. Buzz, who was the second person to set foot on the moon in 1969
10. Of the moon
13. Particular way of standing
14. Go around, as the moon does with respect to Earth
15. Touch
18. Agcy. involved in the Apollo Moon landing missions
20. ½
22. Saturn's largest moon
23. Motivation

DOWN

2. Region
3. Scientific inquiry
4. Regular, repeating series of events, such as the phases of the moon;
5. "Lunar _____," which occurs when the moon passes into the Earth's shadow
6. Body orbiting a planet
8. Goddess associated with the moon
11. Stone

12. The high and low of these are caused by the moon
16. Roman equivalent of Artemis
17. Allegories
18. When the moon is visible
19. "Eureka!"
21. Phase of the moon in which it appears completely illuminated

Answer on page 317

ACROSS

2. Gender of Mars, the god of war
4. Persistence; also, the name of NASA's rover on Mars
6. Percentage, frequency
7. Mars is primarily made up of carbon _____
9. Collection
10. Pinkish
11. Colorless, odorless, unreactive gas, chemical symbol N;
15. Expedition into space
19. Rate of speed in a certain direction
21. Observe
22. Opposite of "entry"

DOWN

1. Crib
2. The nearest planet to the Sun
3. "Solar _____," an alternative to chemical fuels
4. Orbiting celestial bodies
5. Color used to describe Mars
8. Movement around the Sun
9. Weather event
12. NASA's _____ missions to Mars; named after Scandinavian explorers
13. Extraterrestrial vehicle for driving over rough terrain, such as Perseverance on Mars
14. Of the Sun
16. Glacial
17. The gas needed for breathing
18. Large body of water
20. Clock

Answer on page 318

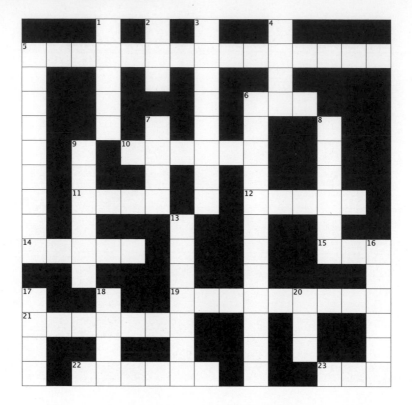

ACROSS

5. Medicinal drugs
6. Tissue layer
10. Eye part
11. Extent of damage
12. Lead or gold, for example
14. Treatments, antidotes
15. Age
19. Pain-killing, pain-numbing
21. Long-lasting, like a disease
22. Medical treatment
23. Creative enterprise

DOWN

1. Pharmaceuticals
2. Inoculation, slangly
3. What the doctor prescribes; also, the general name of the doctor's profession
4. Tablet, capsule
5. Substance stimulating the growth of microorganisms to health promote health, especially in the intestines
6. Branch of medicine dealing with drugs
7. To restore to health
8. Medicinal amount
9. Often compared with "nurture"
13. Digestive aid
16. Opposite of reject
17. Mark on the skin left by a wound
18. The femur, for one
20. Organ with a hammer

Answer on page 318

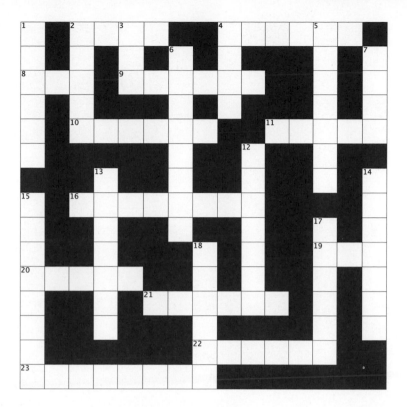

ACROSS
- **2.** Opposite of strong
- **4.** Cat
- **8.** Opposite of in
- **9.** Movement
- **10.** Automatic reaction
- **11.** Velocity, which a force can alter
- **16.** The force resisting the motion of material elements sliding against each other
- **20.** Declaration, demand
- **21.** Opposite of weak
- **22.** Sir Isaac, English scientist who studied forces in an in-depth manner
- **23.** Fastening

DOWN
- **1.** Kind of energy or number
- **2.** Need for life
- **3.** Goal, objective
- **4.** Movement of water in a current or stream
- **5.** Type of power which can be obtained via fusion or fission
- **6.** Might
- **7.** Capacity
- **12.** Pressure
- **13.** The force that keeps objects from floating off the Earth
- **14.** Draw
- **15.** Charged
- **17.** Fixed amount
- **18.** Coil

Answer on page 318

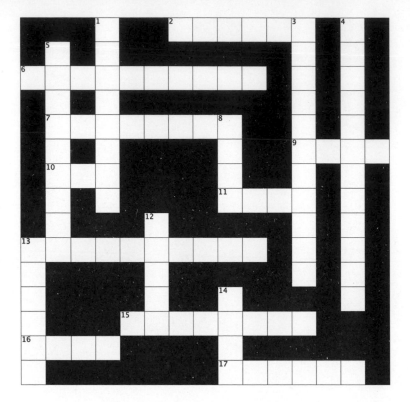

ACROSS

2. "The Last ____," one of da Vinci's masterpieces

6. Da Vinci invention often found in math class

7. The science or study of aircraft, to which da Vinci contributed before the age of airplanes

9. Seasoning spilled in da Vinci's *Last Supper* painting

10. Color shade or tint

11. "Mona ____," called *La Gioconda* (the playful one) in Italian; another da Vinci masterpiece

13. Aircraft designed da Vinci centuries before air travel

15. Da Vinci was born in a little town just outside this major Renaissance city, sometimes called the cradle of the Renaissance

16. Armored vehicle; another of da Vinci's designs

17. Aeronautical

DOWN

1. Many devices built by da Vinci

3. Era in which da Vinci lived

4. One of da Vinci's rivals in art; he sculpted the *David*

5. Device invented by da Vinci that uses air to slow an object

8. It is used to hammer something down

12. Inventor's construction

13. Well-being

14. Identity

Answer on page 318

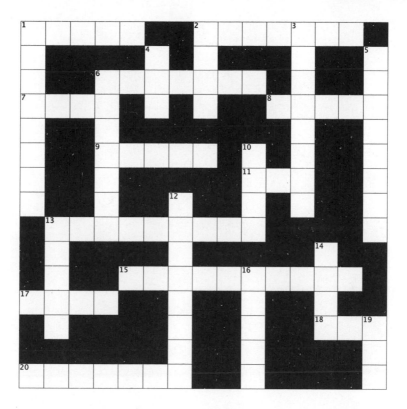

ACROSS

1. Decomposition
2. Out of one hundred
6. Odometer measure
7. The "T" in D = R ´ T
8. Stopwatch
9. Richter _____
13. Rate of something that occurs repeatedly
15. Size of something in relation to something else
17. Warm
18. Shred
20. Swiftly

DOWN

1. Space
2. Design
3. Approximation
4. Everything
5. Mutable
6. Standard unit for describing size or amount
9. Each, as in "_____ person,"
10. A lot
12. Rate of speed
13. Stationary
14. Sixty minutes
16. Percentage, proportion
19. Each

Answer on page 318

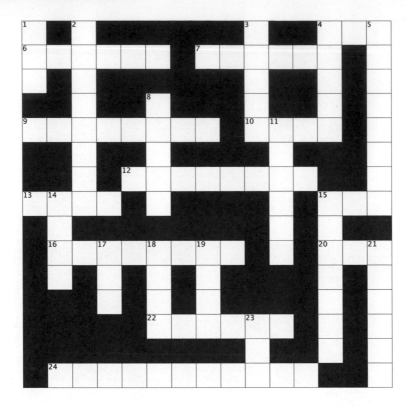

ACROSS

6. What aviation is about

7. Name of the brothers who built the first successful motor-operated airplane

9. Blimps

10. Structure at the rear of an aircraft providing stability during flight

12. Type of aircraft designed for use in battle

13. Toy flown in the air with a long string

15. Place

16. An aircraft's main body section

20. Opposite of old

22. Force of a jet or rocket engine that moves an aircraft in the direction of the motion

24. Aircrafts propelled by sets of horizontally revolving overhead rotors

DOWN

1. Ship for ETs

2. Machine capable of flight

3. Aviator

4. Reach a condition where aircraft speed is too low to allow for its effective operation

5. Aerospace engineering

8. Public

11. Recognized flight path or route

14. The scoop

15. The opposite of take-off

17. Where aircraft fly

18. The airplane's rise from the ground during take-off

19. Landing ____, touches the ground first while landing and leaves it last during take-off

21. Load, cargo

23. Take in

Answer on page 318

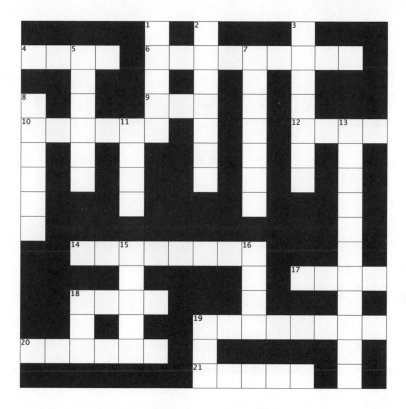

ACROSS

4. Keep
6. Angles with the same shape and size; mirror images of each other
9. It is equal to 2π radians; a letter in the Greek alphabet
10. Angle that is more than 90° and less than 180°
12. Opposite of odd
14. An angle of 180°
17. Many
18. Side
19. The branch of mathematics that deals with angles
20. Sixtieth of a degree of angular measurement
21. Identical

DOWN

1. An angle less than 90°
2. Pointed
3. Units of angle measurement
5. The point where two lines meet to form an angle
7. Alternative angle measures to degrees; they are equal to an angle at the center of a circle whose arc is equal in length to the radius
8. Trigonometric function, which expresses the ratio of the side adjacent to an acute angle to the hypotenuse
11. Trigonometric function, which expresses the ratio of the side opposite a given angle to the hypotenuse
13. Triangle with its three equal sides and angles
15. Type of angle equal to 90°
18. Indefinite period of geological time
19. MBA seeker's exam

Answer on page 319

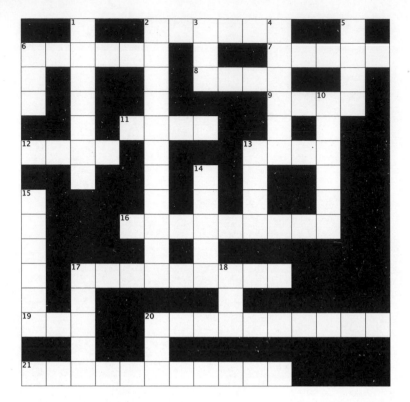

ACROSS

2. Measurement system used in most of the world
6. Figure made up of four equal sides
7. Suggest, infer
8. Assert (as fact)
9. Shape that is used to make ice cream wafers
11. Shape of a box
12. Curved structure supporting the weight of something above it
13. Captain
16. The boundary of a closed geometric figure or shape
17. A quadrilateral with 90° angles, but not a square
20. Unit of energy
21. Three-sided
22. Technique in the calculus for determining the area under a curve

DOWN

1. _____ area, the area of an outer part of a solid such as a cube
2. The size of something
3. "Thanks in advance" in Internet slang
4. Shape whose area is πr²
5. Piece of evidence collected at a crime scene
6. Test taken in HS
10. Digit
13. Place of residence
14. Straighten
15. Orb, globe
17. Type of angle found in rectangles
18. Transcript stat
19. Theme of this puzzle
21. Asphalt

Answer on page 319

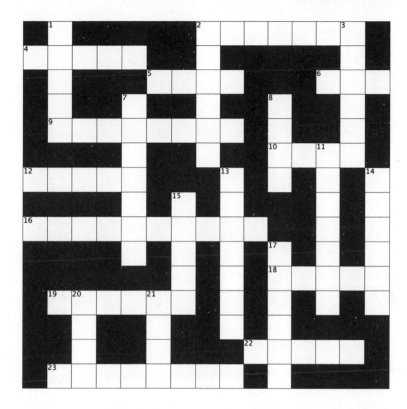

ACROSS

2. Fluids
4. Two pints
5. Ice cream truck option
6. Beer
9. Volume
10. Measures of capacity used in cooking
12. Sum; aggregate
16. Measurable extent
18. Solid geometrical figure that refracts light
19. Cylindrical glass container used typically in laboratories
22. Eye medication dispensers
23. Closed solid with two circular bases connected by a curved surface; its volume is: $V = \pi r^2 h$

DOWN

1. What the exponent in "n^3" refers to
2. Measurement needed to calculate volume of a cube
3. Not liquids or gasses
7. Units of liquid capacity used for measuring gasoline
8. Large amount
11. A triangular solid; the structure that describes some ancient structures, such as those in Egypt
13. Degree of compactness, defined as mass divided by volume
14. The fluid part of blood
15. Unit of measurement equal to 1.057 quarts
17. Shape of the Earth
20. Simple
21. Not odd

Answer on page 319

ACROSS

1. Steep slope separating areas of land at different heights; the bottom of a cliff
3. Knight's title
5. Small stones
7. American space agcy.
9. Building where grain is ground into flour
10. Rock composed mainly of quartz
11. Rock formed by solidification of lava or magma
14. Rock formed from sediment of water or air
16. Blend, combine
17. Strata of rocks
20. It erupts and spews lava
21. Rocks, pebbles

DOWN

2. A route through the mountains
3. Oil source
4. This puzzle's theme
6. Large rocks smoothed out by erosion
8. Hard, crystalline limestone
9. Extremely hot liquid under Earth's surface that is called lava when it erupts onto the surface
10. One of the three states of matter
12. Object with cultural value
13. Wearing away of sediment
15. Membrane behind the cornea of the eye
17. Everything
19. Dashed

Answer on page 319

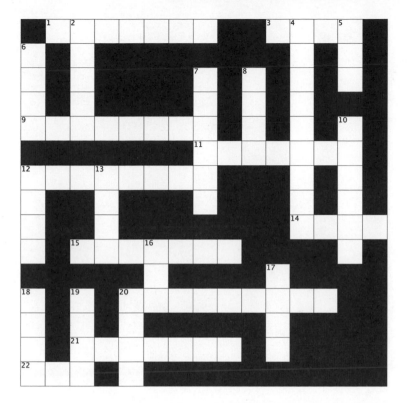

ACROSS

1. Headache medication
3. Often contrasted chemically with "base"
9. Compound containing two oxygen atoms in its molecule; "hydrogen ____," is used to prevent infection from minor skin cuts or burns
11. A colorless gas with a pungent smell used as a cleaner for mirrors and glass
12. Kind of acid used in the manufacture of fertilizers, pigments, dyes, drugs, explosives, detergents, and inorganic salts
14. Stretches of geological time
15. Volatile liquid present in coal tar and petroleum
20. Like some deadly chemicals
21. Corrosive
22. Rodent that looks like a large mouse

DOWN

2. Used to sweeten coffee or tea
4. Formed by reaction of carbon dioxide with a base; "calcium ____" is used as an antacid against heartburn
5. It adds a color to something
6. Fall
7. Common name for a solution of sodium hypochlorite or hydrogen peroxide; used to whiten or sterilize substances
8. Froth; mass of bubbles
10. "____ soda," technically, sodium bicarbonate, used in cooking or cleaning
12. A bar of this is used for cleaning
13. Satisfactory
16. Animal park
17. Carbonated water
18. Opposite of "under"
19. A real piece of information
20. Addition

Answer on page 319

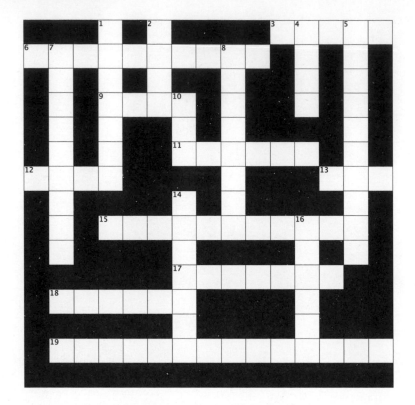

ACROSS

3. Bait

7. Study of the forms of living things; in linguistics it refers to the structure of words

9. Building block of all living things

11. The science of plants

12. Tears, cuts

13. Large deer native to North America

15. Tendency of living organisms to maintain a relatively stable equilibrium

17. Ordain

18. Scientist Charles who put forth the theory of evolution and is sometimes called the "father of modern biology"

19. Process that involves chlorophyll and generates oxygen; in green plants the process converts sunlight into chemical energy

DOWN

1. Darwin's "On the Origin of ____" (1859)

2. Upper layer of earth where plants grow

4. Egg cell

5. Set of life-sustaining chemical processes in organisms

7. Life forms; living things

8. The study of heredity

10. Research facility

14. Lions and tigers, for example

16. Containing salt

Answer on page 319

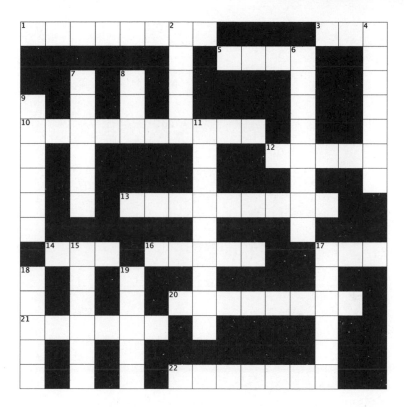

ACROSS

1. The science of heredity
3. Bone that is part of a "cage," which protects the thoracic cavity
5. Structure, shape
10. Evolutionary adjustment
12. What bees, wasps, ants, and scorpions can do
13. "Natural _____," natural processes which determine which organisms thrive better than others
14. Staff
16. The "Scopes _____," famous for its defense of the theory of evolution
17. Spherical green seed, whose plant was used by Mendel to develop the science of genetics
20. Porous
21. Warm blooded vertebrate animal, including cats, dogs, and humans
22. "On the Origin of _____" (1959), Darwin's key book

DOWN

2. Table that organizes information
4. The study of living organisms
6. Change, sometimes in the structure of genes
7. Genetically-based characteristics
8. Strong, robust
9. Charles, who was the proponent of the theory of natural selection
11. Passed on genetically
15. Living organism, distinct from plants
17. Trees, flowers, shrubs, and ferns, for example
18. An animal with one or two humps
19. Have in common

Answer on page 320

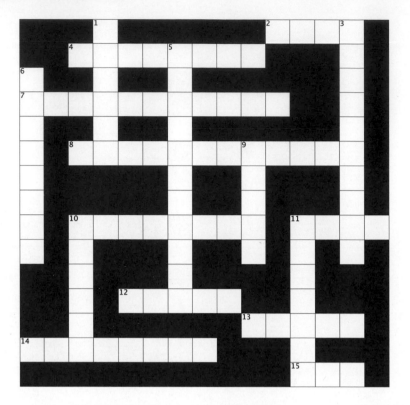

ACROSS

2. Hive dwellers
4. The passing on of genetic characteristics genetically from one generation to another
7. DNA molecules carrying genetic information
8. Process of producing offspring
10. It results from a change in genetic structure
11. Basic unit of heredity
12. They make up the skeleton in humans
13. Permitted
14. Mobile phone system
15. Deoxyribonucleic acid

DOWN

1. Term referring to the genetic material in a cell or organism
3. Charting the order in which amino acids are arranged in a protein
5. Inherent qualities of character or mind
6. Type of characteristic not inherited, but resulting from environmental effects
9. An organic structure grown from a single cell and genetically identical to it
10. Gregor Johann _____ (1822-1884), called the "father of genetics" after he demonstrated the transmission of characteristics in peas in a predictable way
11. Roared

Answer on page 320

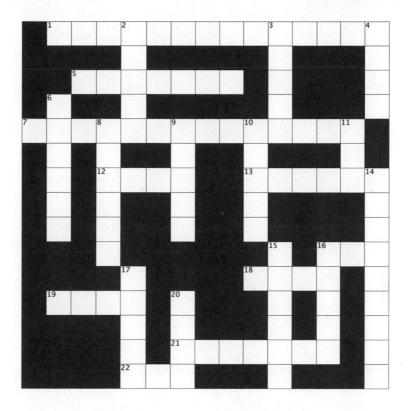

ACROSS

1. Plant process that produces oxygen
5. Organisms with shared traits
7. Organization in categories
12. It falls from clouds
13. Development, maturation
16. Prefix attached to words to indicate a lower level or position
18. Part that attaches a plant underground, providing water to the rest of the plan
19. Body of a plant usually above ground, which supports flowers and leaves
21. Roses, tulips, and geraniums, for example
22. Untruth

DOWN

2. Plants with a trunk, leaves, and branches
3. Emotion
4. The plant's unit of reproduction
6. Living organisms like trees, flowers, and shrubs
8. Woody plants smaller than a tree with several stems arising from the ground
9. Spore-producing organisms that include molds and mushrooms
10. Aquatic organisms that have the ability to perform photosynthesis
11. Fruit with a hard shell that contains a kernel
14. An animal that feeds on plants
15. Protein food for the honeybee colony
16. Ends
17. Sniff
20. Existence, being

Answer on page 320

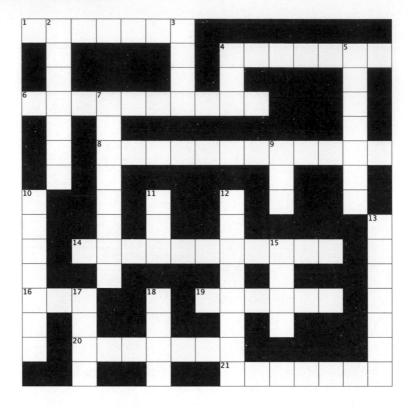

ACROSS

1. Class of animals that includes humans, felines, and canines

4. Birds able to mimic the human voice

6. Branch of biology studying pre-birth development

8. Extinct winged reptiles of the late Jurassic period

14. Animals with a backbone or spinal column, including mammals, birds, reptiles, and fishes

17. Fisher's need

19. Hoofed domesticated mammals with a mane and tail

20. Group of organisms with shared characteristics

21. Horses or other members of the horse family

DOWN

2. Living organisms differentiated from plants

3. Aquatic mammal with feet developed as flippers, which returns to land to breed

4. Omnivorous hoofed mammal with a flat snout for rooting in the soil

5. Slow-moving reptiles, with a domed shell into which they can retract

7. Animals such as snakes, turtles, and lizards

9. Org. responsible for public health

10. Animals such as tigers, lions, and everyday cats

11. Animal doc

12. Land-dwelling reptile with a hard shell

13. Animals such as mice, rats, and hamsters

17. Job

18. Group of wolves

Answer on page 320

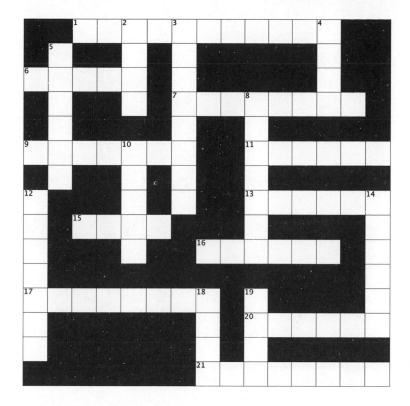

ACROSS

1. There are two of these in the brain—a left and a right one

6. The parts of the cerebrum of the brain; the main ones are called: frontal, parietal, occipital, and temporal

7. Mass of gray matter inside each cerebral hemisphere that involves the experience of emotions; it is part of the limbic system

9. Complex bodily system that coordinates its actions and sensory information

11. Nerve cell

13. System of nerves near the edge of the cortex concerned with instinct, mood, emotions, and drives

15. "_____ matter," refers the tissue of the brain and spinal cord, consisting mainly of nerve cell bodies; it is compared with *white matter*

16. Organs that secrete chemical substances for use in the body or for discharge into the environment

17. Using the brain to gain understanding

20. _____ of Miletus (c. 624–545 BCE), Greek philosopher, considered by Aristotle to be the founder of physical science; he was one of the first to explain physical phenomena scientifically

21. One of the four main brain lobes

DOWN

2. Quantity of matter

3. Junctions between nerve cells across which electrical nerve impulses pass

4. This spiritual entity was once considered to be located in the brain; regarded as immortal

5. The outer layer of the cerebrum, composed of gray matter

8. "Basal _____," structures linked to the thalamus and involved in coordination of movement.

10. A biological structure, such as the brain

12. One of the four main brain lobes

14. "_____ callosum," the cluster of nerve fibers joining the two hemispheres of the brain

18. Present

19. Central trunk of the brain

Answer on page 320

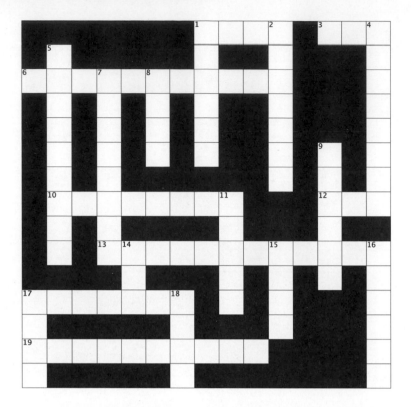

ACROSS
1. Location of the brain
2. Number of eyes (and ears, hands, or feet)
6. The Greek founder of medicine; he established an oath stating the obligations and proper conduct of doctors, still recited in medical schools today
10. Signs indicating the condition of disease
12. Spring month considered to be the National Physiotherapy Month by some practitioners
13. Cancer-causing
17. Medical operation
19. Branch of medicine concerned with the health and diseases of the heart

DOWN
1. Well-being
2. Illness, sickness
4. Branch of medicine dealing with cancer
5. Identification of a disease by examination of symptoms
7. Disease prevalent over a whole country or the world; as, for example, COVID-19
8. It defeats a disease
9. Resistant to a particular infection or toxin because of the presence of specific antibodies
11. Nasal cavity
14. Era
15. DNA carrier
16. Serious disease caused by the uncontrolled division of abnormal cells
17. Opposite of healthy
18. The Medical School of this university, in New Haven, Connecticut, is considered one of the best in the US

Answer on page 320

ACROSS

3. Organ located in the abdomen behind the stomach which functions as a gland, which helps break down proteins, fats, and carbohydrates
7. Cluster of similar cells that carries out a specific function
9. Located in the upper right abdomen, it is a glandular organ that excretes bile
10. Organ that's a symbol for love
11. The olfactory organ
13. Bean-shaped organ behind the abdomen, which acts to filter blood to create urine
14. Ear parts
16. Blood vessels
19. Main part of the large intestine
20. Organ located behind the stomach
22. _____ and now
23. The tasting and licking organ

DOWN

1. Organs that absorb nutrients and can be upwards of 15ft long
2. System that coordinates actions and sensory information
4. The organ where the digestion of food occurs
5. Breathing organs
6. Some species of octopus have this color blood
8. Backbone
12. Looks at
15. Organ that stores urine
17. Nothing, zero
18. Oral cavity
21. God of Greek mythology with a man's body and goat legs

Answer on page 321

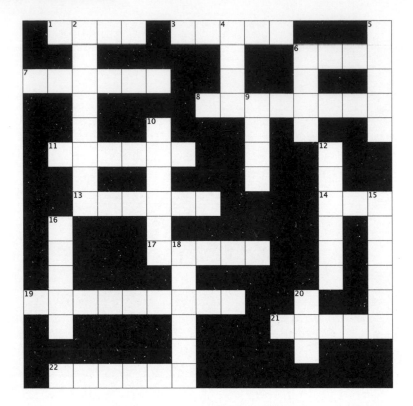

ACROSS

1. The human form
3. Body parts containing marrow
6. Eye drop
7. Structures such as the heart and the liver
8. The body's framework, consisting of bones, cartilage, and ligaments
11. Flesh, for example
13. Bundle of fibrous tissue
14. One of the two walking limbs
17. The arms and the legs
19. It connects the throat to the stomach
21. They allow us to hold things and to write, among other things
22. It is in the neck and might hurt when someone contracts a cold or the flu

DOWN

2. Living thing
4. It connects the head to the body
5. Torso, upper body
6. Pedicurists concerns
9. Look sat
10. Of the mouth
12. Hip location
15. Organs secreting chemical substances
16. Upper body, breast
18. Swallow
20. Mandible

Answer on page 321

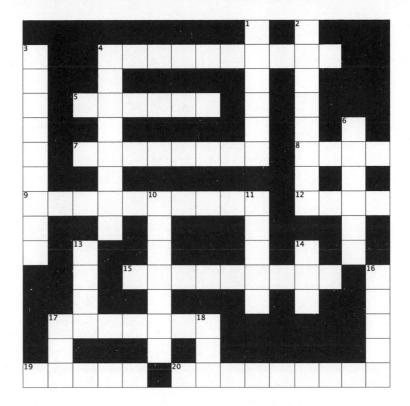

ACROSS

4. Serious infection caused by bacteria that make toxin; symptoms include breathing difficulties, heart failure, and paralysis; now rare because of immunization

5. Examples of this diseases include leukemia and sarcoma

7. The study of heredity

8. It causes an urge to scratch

9. They can be bacterial or viral

12. Looks at

15. Noncontagious skin disease characterized by red, very itchy patches

17. Infectious childhood viral disease marked by fever and a red rash on the skin; known medically as rubeola

19. Intensely itchy skin rash, with red welts, caused

typically by an allergic reaction

20. Viral infection that attacks the respiratory system

DOWN

1. Viral disease affecting the skin or the nervous system

2. Throbbing headache, often accompanied by nausea

3. These include hay fever and reactions to pollen (such as

sneezing and tearing)

6. Skin inflammation, with blisters causing intense itching and bleeding

10. Two small masses of lymphoid tissue in the throat

11. Condensation, water vapor

13. Pimples

14. Twitch

16. Suppository

17. Hosp. scan

18. Celestial body that provides heat and energy

Answer on page 321

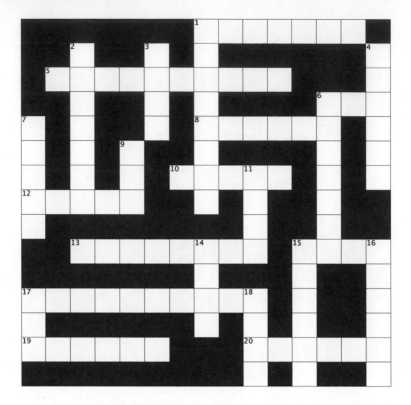

ACROSS
1. Eye specialist, ophthalmologist
5. Lenses that improve vision
6. Yoga need
8. Layer forming the front of the eye
10. Surgery using "amplified light beam" technology
12. Relating to vision or the eyes
13. Places for pupils
15. Eye part that surrounds the pupil
17. Vision correctors
19. Openings in the center of the iris which change size to regulate the amount of light reaching the retina
20. Eyesight

DOWN
1. Lens maker
2. "_____ lenses," are corrective lenses placed on the eyes
3. Part of the eye
4. Layer at the back of the eye where images are formed
6. Part of the retina at the back of the eye
7. Fluid inside the eyes
9. Curved segment, semicircle
11. The organs of vision
14. "Vision _____," which can be caused by macular degeneration
15. Force something on someone
16. Hard look
17. Paranormal abilities, briefly
18. Keep

Answer on page 321

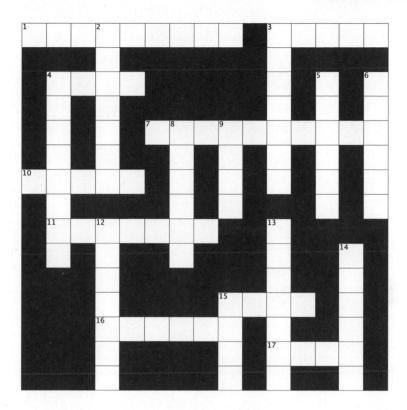

ACROSS

1. Living things
3. A dry flake of skin
4. Masculine
7. Units of heredity
8. Organized forms
10. Delete
11. Central part of an atom
14. "Blood _____," Coagulated blood
15. Willing
16. Protected through vaccination
17. Tidy

DOWN

2. Genes
3. Of the body, corporeal
4. Thin layer of a cell; it can shield it
5. Ongoing story
6. Fibrous tissue
8. Cluster of cells carrying out a specific function
9. Single element
12. Asexual reproduction
13. Chemicals secreted directly into the blood; an example is adrenaline
14. Fertilized ovum
15. DNA part

Answer on page 321

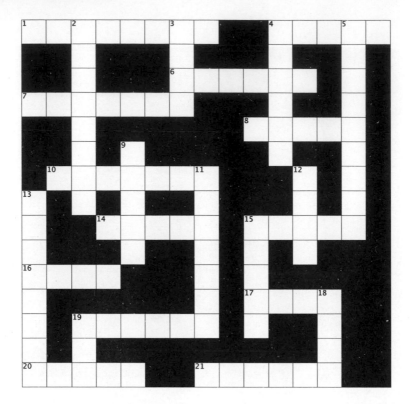

ACROSS

1. Category that includes snakes, lizards, and tortoises
4. Large animal groups, such as cattle or elephants
6. Arboreal primates with a long tail, found only in Madagascar
7. Warm-blooded vertebrates distinguished by the possession of hair or fur; it includes humans, dogs, and cats
8. Aquatic mammals with flippers
10. Category that includes cats, lions, and tigers
14. Serpent
15. Groups of wolves
16. Small horse
17. Animal with antlers
19. Groups of birds
20. Large feline with a yellow-brown coat and black stripes, native to Asia but on the verge of extinction
22. Large marine animals with a hole on the head for breathing; a white one was described by Melville in *Moby Dick*

DOWN

2. Category that includes humans, chimpanzees, and gorillas
3. Snakelike, slippery fish, which seem to give off light
4. Equines
5. Extinct prehistoric reptiles
9. Kings of the jingle
11. Organisms with a shared evolution
12. Small rodents that might invade the home
13. Large mammal with a characteristic long trunk
15. A group of lions
18. Rodents larger than mice, which might invade human habitations in search of food
19. Soft, pear-shaped fruit with many small seeds

Answer on page 321

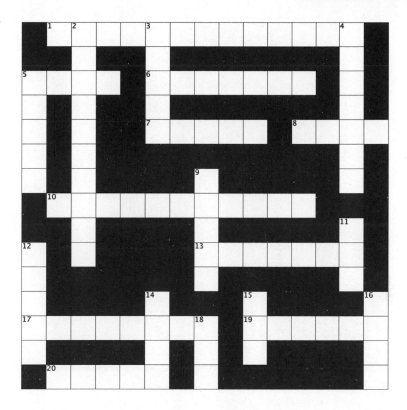

ACROSS

1. Process whereby a caterpillar becomes a butterfly
5. Small insects that live in colonies; some are called "fire" or "carpenter"
6. Insects that make mounds of cemented earth and feed on wood
7. Malice
8. Wingless jumping insect which feeds on the blood animals, such as cats and dogs
10. Beetle-like pests found typically in habitations
13. "_____ mantis"
17. Hot weather pest that can bring diseases such as the West Nile virus
19. Insect forms, between an egg and a pupa
20. "_____ widow spider," a highly venomous spider

DOWN

2. The science of insects
3. Insects related to butterflies, and chiefly nocturnal
4. This category of insects incudes tarantulas and black widows
5. Sap-sucking insects called commonly greenflies or blackflies
9. Small winged insects sometimes mistaken for bees
11. Number of years that a person has lived
12. Garden helpers
14. Parasitic arachnid that can pass on Lyme disease by sucking the blood of its host
15. Common household annoying insect
16. Hive workers

Answer on page 322

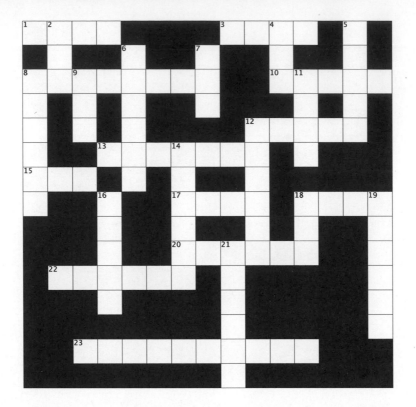

ACROSS

1. Large mammal that walks on its feet and has thick fur
3. Animal that roars
8. Arachnid with pincers and a poisonous stinger at the end of tail, which it can curve
10. Common insects
12. They can cause Lyme disease
13. Creeping insects which can also fly; the "longhorn" ones have long antennae
15. Snakelike fish, which is rather slippery
17. Animal closely related to sheep
18. They make honey
20. A mouse or rat, for example
22. Animals seen on safari
23. Dangerous, semi-aquatic reptiles

DOWN

2. Comp. key
4. Opposite of on
5. They are found commonly on dogs and cats
6. Tarantula, for example
7. Colony-living insect; known for its industriousness
8. Limbless reptiles, some of which are quite venomous
9. A member of an imaginary race of malevolent humanlike creatures
11. Small parasitic insects that live on the skin of mammals
12. Dangerous fly
14. Black-striped large cats
16. Attention-drawing action
18. Cave dweller
19. Long-bodied marine fish, mainly predatory
21. "Rubber _____," a toy that looks like this webbed-footed animal

Answer on page 322

ACROSS

1. Microorganisms, such as bacteria
5. Mouth lesions caused typically by the common cold
7. Not old
8. COVID-19 for example
10. Hip
11. Respiratory virus
14. Serious inflammation of the lungs caused by bacterial or viral infection
15. Common sexually transmitted disease
17. Contagious viral disease marked by the swelling of the glands located near the ears

DOWN

2. Diseases caused by germs
3. Name for a hemorrhagic fever virus that leads to internal bleeding; found mainly in Africa
4. Transfer of viral or bacterial DNA from one cell to another
5. Leaves a mark
6. Cell in which a virus multiplies
9. Infectious bacterial disease of the small intestine, contracted from infected water supplies, causing vomiting and diarrhea
12. Bloodsucking parasite
13. Regions, areas
16. Edible climbing plant, found typically in tropical and subtropical countries

Answer on page 322

ACROSS

3. They are made up of DNA and carry genetic information

7. Ribonucleic acid (*abbr.*); it carries information from DNA for controlling the synthesis of proteins

8. Research site

9. "DNA _____," molecular-weight size marker

12. Genetic instructions

14. Term referring to the complete set of genetic information in an organism; stored in chromosomes

16. Duos

17. They are essential to building muscle mass

20. DNA molecules are composed of two polynucleotide _____ that coil around each other, forming a double helix

DOWN

1. "_____ helix," a phrase that describes the appearance of the DNA

2. Analysis of a person's psychological and behavioral characteristics

4. It is "double" in the structure of the DNA

5. Bud, fertile spore

6. Make up of a DNA strand

10. Test

11. Living things

13. Related to criminal investigations

15. "Genetic _____," used to show the relative locations of genes

18. Streak

19. Not well

Answer on page 322

ACROSS

3. Disease-causing bacteria
6. Single-celled microscopic organisms, including amoebas and ciliates
7. Common term for bacteria, colloquially, "bugs"
8. Hair-like projections from some kinds of cells
9. Infectious agent
12. Disease-causing microorganism
13. A single-celled organism, commonly found in pond water and ditches
16. The science of microorganisms
17. Seaweed, for example
18. A period of difficulty or danger; also, the turning point of a disease

DOWN

1. Commonly called "germs"
2. Lab instrument that allows scientists to get a look at microbes
4. Single-celled organism that includes protozoa and algae
5. Organism that lives in or on another organism, deriving nutrients at the expense of the host organism
10. Salt Lake City's state
11. Microscopic fungus consisting of single oval cells, often used in baked goods
14. The heart or liver, for example
15. Skin structures that secrete sweat
16. Penicillin source

Answer on page 322

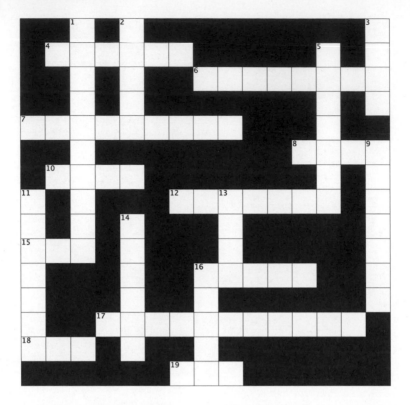

ACROSS

4. Name of a group of mice living together

6. Burrowing rodents with large cheeks and pouches, often kept as pets

7. Tree-dwelling rodents with a bushy tail, which feed on nuts and seeds

8. Sense

10. Small rodents that often invade human habitations

12. Animal category that includes apes and humans

15. Yellowish discharge from infected tissue

16. Canines, for example

17. Like animals that feed on plants

18. Body of water, such as the Mediterranean

19. _____ Today

DOWN

1. Large rodents with defensive quills on their bodies, which they can shoot at predators

2. Spectrum part

3. Opposite of fall

5. Broad-tailed rodents that is gnaw through tree bark in order to feed themselves and build dams

9. Small rodent with a short tail, found in the Arctic tundra

11. Burrowing rodents with fur-lined pouches on their cheeks

13. Stag or buck

14. "_____ pig," Tailless rodent kept usually for lab work

16. Appendages that most rodents have

Answer on page 322

ACROSS

1. Plant with a trunk and leaves
2. Bushes or hedges
6. They bloom in Spring
7. Plants used for flavoring, food, or for medicinal purposes
11. A place built for vegetables, trees, flowers or other plants
12. Concern
13. Flowerless green plant that grows in damp habitats and reproduces by means of spores
14. Community-supported agriculture (*abbr.*)
16. Parts of the plants that are underground
17. It comes out at night in the sky, influencing the growth of plants
19. Substance used to destroy plants
20. Unit of lettuce or cabbage
21. Fiber found in plants; in flowers it is the male reproductive part

DOWN

1. Clip, as a hedge
3. The practice of garden cultivation
4. These are planted in the ground for plants to be born
5. Plant
6. "_____ and fauna,"
8. Nursery purchases
9. Flowerless plants
10. Yearly
11. It covers lawns and gardens
14. The four-leaf kind of *this* is believed to bring good luck
15. Leaves
18. Wild plants that are unwanted in a garden

Answer on page 323

ACROSS
3. The study of fungi
6. Potato
7. Lethal
10. Heat-generating star
11. One-celled unit of reproduction, including for fungi
12. Mushrooms' tops
13. Doctor's order
14. Fungus considered a culinary delicacy
17. Illness
18. Thin membrane on a mushroom, or a bridal accessory

DOWN
1. Minute fungi occurring typically in warm conditions, especially on organic matter
2. Spore-bearing parts of a fungus, in the form of a rounded cap, generally poisonous
3. An edible fungus, with some exceptions
4. Microscopic fungus that causes infections
5. Disease caused by fungal infection, such as ringworm
8. Chemical that destroys fungi
9. Fit to consume
15. Plug
16. Biblical garden

Answer on page 323

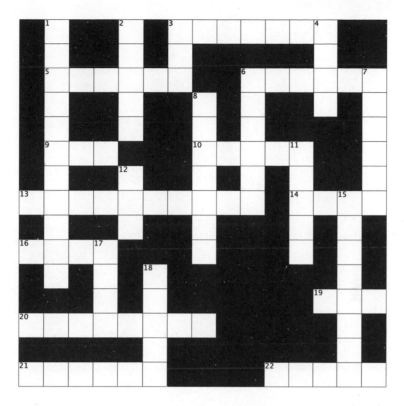

ACROSS

3. Animals with broad tails that build dams
5. Mouse or rat, for example
6. Wild carnivorous mammals of the dog family, which move in packs
9. "_____ constrictor," a big snake
10. Small, burrowing insect-eating mammals with a long muzzle and small eyes
13. Mammals such as kangaroos and opossums
14. Large rodents
16. Mammals with antlers
19. A common feline
20. Nocturnal mammal with a tubular snout, which feeds on ants and termites
21. Marine mammal with two downward-pointing tusks
23. Killer whales

DOWN

1. Animals, including mammals, possessing a backbone
2. Only mammals have these glands
3. Nocturnal mammal with membranous wings
4. Preserve
6. Large marine mammal, found in *Moby Dick* or *Pinocchio*
7. Arboreal weasel-like mammals, with a short tail and brown fur, native to Siberia and Japan
8. Milk-secreting organs of female mammals
11. Small mouse-like insect-eating mammal with a long snout and tiny eyes; also used in Shakespeare's "The Taming of the _____"
12. Body hair on certain animals, from which clothing is sometimes made
15. A state in Australia that has the marsupial called a "devil"
17. Street
18. Aquatic animals which eat fish and come to land to breed

Answer on page 323

ACROSS

7. Bird cage

9. Medium-sized long-tailed bird, mentioned in a famous novel title

10. Bird's call or coo

12. Like doves, they belong to the same Columbidae family; they have small heads and short legs

14. In some regions, robins and wrens are sign that this season is coming

15. Opposite of front

16. What birds lay

18. Bird that famously puts its head in the sand

22. The scientific study of birds

25. Birdsong

26. Perching birds with black plumage and a raucous voice; often associated with bad omens in folklore

DOWN

1. Bird plumage

2. "Blue ____," common North American bird with a blue crest

3. Color of crows

4. The flying forelimbs in birds

5. The action of the beak to search for food

6. Small short-winged songbirds

8. Post or pillar on which birds can rest

11. They are made by birds to lay their eggs and nurture their newborns

13. It connects the head to the body

15. Gesture requiring attention

17. Male goose

19. "____ of Cancer" or "____ of Capricorn"

20. Large water bird with a long neck and webbed feet

21. Thrush with a reddish breast

23. Plant where you might find a bird's nest

24. Wise birds

25. Avian sounds

Answer on page 323

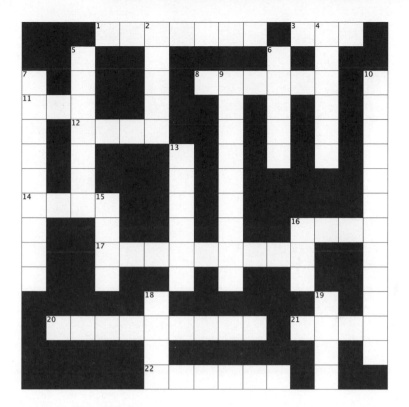

ACROSS

1. Silvery fish that of commercial importance as a food; a red one is said to indicate a deception
3. Fish with a small barbell on its chin; it is a source of liver oil
8. Popular edible fish, with pink flesh, which spawns two or three times
11. Snake-shaped fish, known for its slipperiness
12. Predatory freshwater fish with very large teeth
14. Appendages used by fish to move, steer, or stop
16. Croaking amphibian, like Kermit
17. Reptile found in Florida's Everglades
20. Semiaquatic reptile, often confused with 18-Across
21. Plate
22. People who wear special suits to work underwater

DOWN

2. Increase
4. Elaborate
5. Small toothed whale known for its high intelligence, and use of form of sonar called echolocation
6. Opposite of "big,"
7. Marine animals that sometimes have poisonous stinging tentacles
9. Class of vertebrate animals that comprises frogs, toads, and salamanders
10. The study of the physical and biological aspects of the sea
13. Large edible fish of warm seas that reaches a great weight; distinguished by a spear-like snout; in Ernest Hemingway's 1952 novel *The Old Man and the Sea*
15. Aquatic mammal with flippers
16. Cook in oil or fat
18. A type of frog
19. Respiratory organ of most fishes

Answer on page 323

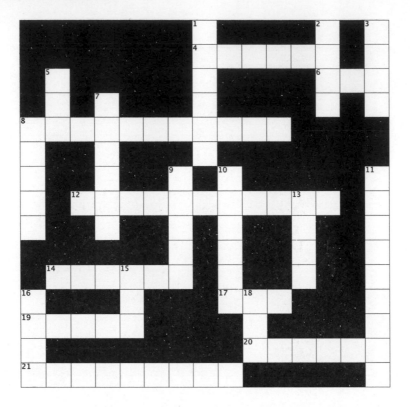

ACROSS

4. Large, arboreal, American lizard with a spiny crest and greenish coloration
6. Nothing
8. Meat-eating
12. Branch of zoology dealing with reptiles and amphibians
14. Taxonomic category above "class" and below "kingdom"
17. Ovum
19. Like a lizard's skin
20. Terrapin
22. Large semiaquatic reptile similar to a crocodile

DOWN

1. Reptile with a long body and tail, movable eyelids, and a scaly, spiny skin; the largest species is called the "Komodo dragon"
2. Opposite of sea
3. _____ - blooded, like reptiles
5. Constrictor
7. Serpents
8. Venomous snake native to Africa and Asia that forms a hood with its neck when disturbed
9. Defense for some reptiles
10. Chameleons use this to catch insects
11. Land-dwelling reptiles with a hard shell
13. "_____ monster," venomous lizard native to the southwestern US
15. Set down
16. Continent that is home to the reticulated python
18. Understand

Answer on page 323

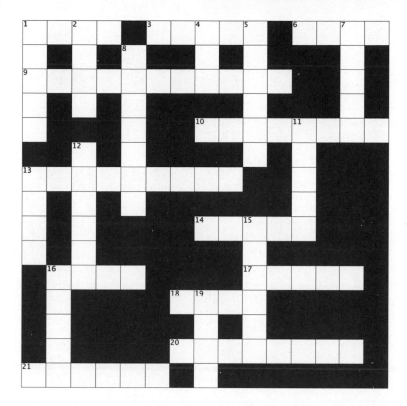

ACROSS

1. Let out
3. "____ cold" is also known as acute bronchitis
6. It connects the head and the body
9. Bodily system that allows animals to breathe
10. Migraine, for example
13. Blocked, like the nose during a cold
14. Aches
16. Opposite of strong
17. High body temperature
18. Skeleton part
20. Type of germ
21. Paroxysm

DOWN

1. Like the flu
2. It can become congested during a cold
4. Consume food
5. It can become red and sore during a cold
7. Sharp expulsion of air; common during colds
8. Hollow cavities that can become infected or congested during a cold
11. Helpers
12. Quick expulsion of air through the nose; often caused by sickness
13. Common ailment
15. Affect with a disease-causing organism
16. Bread option
19. Of the mouth

Answer on page 324

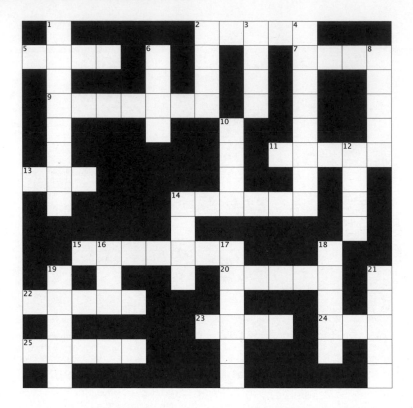

ACROSS

2. Top of the head
5. Smelling organ
7. Of the mouth
9. Tweezer target
11. Vision
13. Mandible
14. They can become rosy when embarrassed
15. Lash sites
20. Light refractor
22. It is surrounded by the lips
23. The back of the neck
24. Falsehood
25. Molars and incisors, for example

DOWN

1. Part of the face above the eyebrows
2. Masticate
3. Shaped like an egg
4. Openings of the nasal cavity that take in air to the lungs
6. Hearing organs
8. Opposite of dark
10. Pimples
12. It grows throughout the body, but especially on top of the head
14. Protruding part below the mouth
16. However
17. Talks
18. Facial expression indicating happiness
19. The facial skeleton consists of fourteen of *these*
21. Uncommon eye color

Answer on page 324

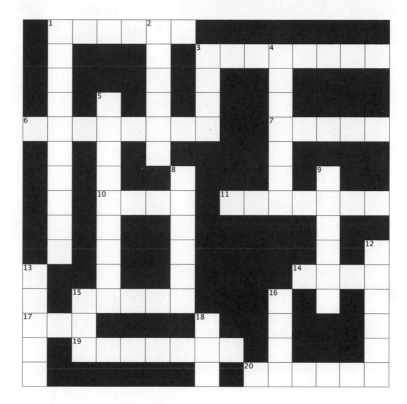

ACROSS

1. Diamond and graphite are its two main forms; its chemical symbol is C
3. Substance formed from two or more elements
6. Synthetic organic materials, such as plastics and resins
7. Weighing balance used in a chemistry lab
10. Thermodynamic energy
11. Wine, beer, or spirits
14. Opposite of acid
15. They turn blue litmus paper red; opposite of bases
17. Motor coach
19. Where cooking is usually carried out
20. "_____ alcohol" is a colorless liquid with an aromatic odor, useful as a solvent

DOWN

1. Salts of carbonic acid; "sodium _____" are called "washing sodas"
2. Quartz and rust, for example
3. Covers
4. Synthetic materials made from organic polymers; used in the manufacture of many household products, including sandwich bags
5. Unnatural
8. Smells
9. Risk
12. "_____ alcohol," or methanol, flammable liquid alcohol originally made by distillation from wood; chemical formula: CH_3OH.
13. "Test _____," which are found commonly in chemistry labs
15. Inquire
16. Identical
18. Number of digits in the decimal system

Answer on page 324

ACROSS

6. They are provided by vaccines, producing specific antibodies

8. Jaw

9. Originate (from)

10. Prescribed quantities of medicine

11. Examples of these include the Bubonic plague, the Spanish flu, and COVID-19

15. They are produced by the immune system in response alien antigens such as bacteria or viruses

17. Vaccination

DOWN

1. Illnesses, maladies

2. Type of disease that results from an aberrant response of the body against its own healthy cells; examples include multiple sclerosis and rheumatoid arthritis

3. Infectious agents that replicate in living cells; they are responsible for such diseases as COVID-19

4. Shortfall

5. Branch of medicine concerned with immune systems and immunization

7. They are injected to get the body to produce antibodies against diseases

12. Reverberation

13. Carrier of genetic info

14. Before all others

16. Midday

Answer on page 324

ACROSS

1. The "fire" and "army" species of these insects are slightly poisonous
6. Noxiousness
7. Beer choice
8. Techniques used in the detection of crimes; these include toxicological analysis
11. Chemical element with symbol Pb (for *plumbum*), that is poisonous if it is absorbed into the body
13. Substance causing someone to vomit
15. Liquid chemical used commonly to whiten or sterilize materials; it is corrosive and dangerous if ingested
18. Opposite of fictitious
20. Consumed
21. Highly toxic protein obtained from the castor-oil plant
22. Opposite of cold

DOWN

1. Medicines that counteract particular poisons; antitoxins, antiserums
2. Antigenic poisons of plant or animal origin
3. Substances capable of causing death, such as cyanide or bleach
4. Attempt
5. Quicksilver
7. Poisonous chemical element with the symbol As
9. The intoxicating constituent in wine and spirits; it can be poisonous in very large amounts
10. "_____ acid," also known as *aqua fortis* (Latin for "strong water"), a highly corrosive mineral acid
12. Raw information, usually numerical
14. Natural material from which minerals can be extracted, such as iron
16. Rested
17. Mane
19. Region

Answer on page 324

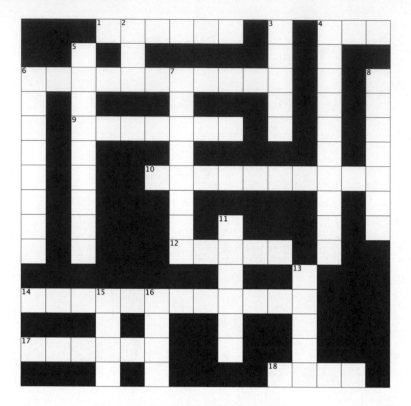

ACROSS

1. Protected, especially after vaccination
5. Org. responsible for public health
7. Any group of RNA viruses that cause a variety of diseases, such as, COVID-19
9. Assails
10. "Social _____," that is, staying away from others
12. Olfactory sense, which can be lost with COVID-19
14. Showing no signs of illness
17. Difficult period of time
18. _____ immunity, which occurs when a population is immune either through vaccination or previous infection

DOWN

2. Game pieces
3. Gustatory sense, which can be lost with COVID-19
4. Infectious
5. Moving air in and out of the lungs, which can be severely affected with COBID-19
6. Expelling air with a hacking sound—a symptom of the coronavirus
8. Extreme tiredness
11. Argue
13. Size, extent
15. Worn on the face to help inhibit the spread of the virus
16. To examine someone for the virus, usually with a swab

Answer on page 324

ACROSS

1. Tiny organisms
5. Branch of biology dealing with cells
7. Liquid part of blood
8. Places for experiments
9. A cell is the basic _____ of organisms
11. Genes mutations
12. "_____ acid," is the general name for DNA and RNA
14. Thin protective layers surrounding cells and surfaces of organs
15. Jelly-like substance that fills cells
19. Optical instruments for viewing cells

DOWN

1. Process in the cells that changes food to energy
2. The basic unit of all living things (what this puzzle is about)
3. Any living thing
4. Specialized structures within a living cell, which store genetic information or assemble proteins
6. Layer surrounding cells
10. Objects
11. Molecule components
13. Substance that provides support for cells
16. Small skin opening that releases sweat
17. Cavity enclosed by a membrane, sometimes filled with fluid
18. Any large tailless primate

Answer on page 325

ACROSS

5. Device used to listen to a patient's heart or breathing
6. Like some masks
7. School email address ending
9. Level
10. Face protectors
12. Controlled environment for the care of premature babies
15. A "health ____" is a health problem affecting populations; an example is a pandemic
16. Prevent from dying
19. Injection of fluid into the bowel to relieve severe constipation
20. Hosp. scan
21. Type of needle needed in medicine

DOWN

1. Used to press down the tongue to allow for inspection of the mouth or throat
2. Respirator
3. Type of surgery used commonly with the eyes
4. Slow injection of medication into a vein or tissue
7. Trouble
11. "High blood ____;" hypertension
13. Device inserted into the bladder to allow urine to drain freely
14. They are worn across the mouth and nose for protection
17. Fluctuate
18. Open hands

Answer on page 325

ACROSS

5. "Natural ____," Darwin's theory that organisms who are most suited to their environments tend to survive and reproduce better
9. Perspective
10. Prefix with center
11. Unit of heredity
12. Process whereby an organism or species becomes better suited to its environment
14. Container
16. Before the present
18. Opposite of under
19. A change in a DNA sequence that leads to evolution
20. Famous trial in which Darwin's theory was challenged

DOWN

1. "HMS ____, name of Darwin's exploration ship and the name of a breed of dog
2. Determined effort under difficulties
3. Natural selection enhances *this* (continuation of existence)
4. Genetic inheritance
6. Darwin's theory
7. Darwin wrote one of the first scientific books on these (moods, urges, etc.)
8. Living things
13. Genetically-determined characteristics
15. Blunder
16. Rate of movement
17. Manage
18. Opposite of in

Answer on page 325

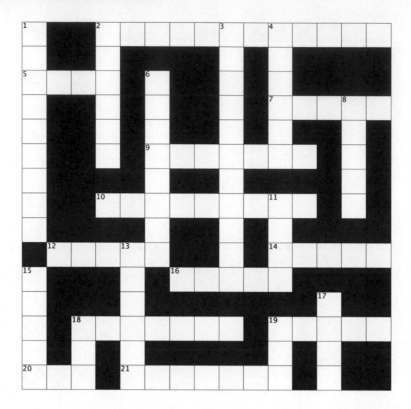

ACROSS

2. Garden cultivation
5. Opposite of sweet
7. Brink
9. Crop yield
10. Animals raised in an agricultural setting
12. Places where crops are grown and livestock kept
14. Crop that's a source of flour
16. "Pine _____," woody fruits of a pine tree
18. Domestic fowl
19. Cultivated and harvested plants
20. Domesticated pig
21. The study how living things interact with each other in their environments

DOWN

1. Farm management
2. Equines
3. Farming
4. They grow on trees and other plants
6. Places to find honeycomb
8. Any cereal crop
11. Bovines
12. Wheat or any other cereal crop
13. Dung
15. Large farm
17. "_____ on the cob"
18. Swine
19. Weep

Answer on page 325

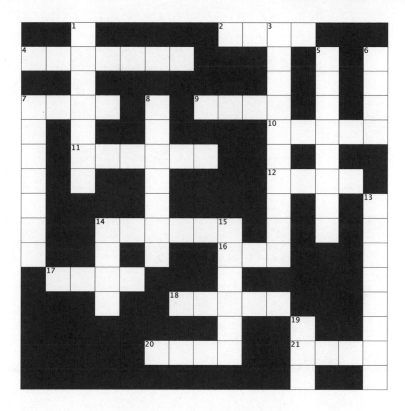

ACROSS
2. Green stone
4. Wearing away of the land from water, wind, and ice
7. Arboreal plant
9. Lumber
10. Parts of trees below the ground conveying nourishment
11. Kindling
12. A wild animal's home
14. They fall from trees in the fall
17. Shrub
18. The woody stem of a tree
20. Trees that produce cones
21. It nourishes trees, falling in drops

DOWN
1. Woodlands
3. The study of trees
5. Tree surgeon
6. In an orchard, what trees are often planted in
7. Arborists tool
8. An organism's natural home
13. Land covered with trees
14. Luxuriant
15. Common Christmas tree
19. Natural material from which minerals can be extracted

Answer on page 325

ACROSS

2. Estrogen and testosterone, for example
4. Looks at
6. Body discharges
7. They are below the knees
8. Part of the skeleton
9. Flesh, for example
10. Hospital need
11. Mandible
13. Back talk
14. Anxious
17. Wasting away of body organs, due to cell degeneration
18. Arms or legs
20. Bed size
21. A group of tissues which contract to generate force

DOWN

1. Animal foot
2. The science of tissues
3. The smelling organ
5. Skin surface
9. Infusion
12. One of the five W's
14. Back of the neck
15. Of the body
16. Relating to birth
18. Ear parts
19. Backbone

Answer on page 325

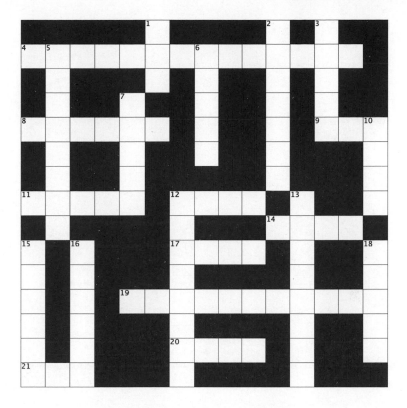

ACROSS

4. Referring to the heart and blood vessels
8. A vein, artery, or capillary
9. Hosp scan
11. Heart pulsations
12. Central part
14. Tablet, capsule
17. Danger
19. Labor pain
20. Exceptional
21. Inquire

DOWN

1. Light-sensitive cell in the retina
2. What the heart should always be doing
3. Part of the heart that allows blood to flow in one direction
5. Tubes through which blood is sent from the heart to all parts of the body
6. The main artery
7. They carry oxygen-depleted blood toward the heart
10. Unit of heredity
12. Relating to the heart
13. Become wider, as in the case of blood vessels
15. Severe pain in the chest, caused by an inadequate blood supply to the heart
16. "Heart _____," common term for coronary thrombosis
18. Regions

Answer on page 326

ACROSS

3. Invertebrate animals of such as insects or crustaceans
6. Parasitic insects that live on mammals and birds
7. Feeling that may lead to vomiting
8. Parasitic arachnids related to ticks
12. Tackle item
13. Org. for doctors
14. Immobile
15. Phylum that includes amoebas, flagellates, and ciliates
17. Swarms
18. Bugs

DOWN

1. Parasites found in the intestines of mammals
2. Bother
3. Parasitic arachnids that attach themselves to the skin, and can cause Lyme disease
4. Frequent loose, watery discharge from the bowels, which can be caused by parasites
5. Serious fever caused by a parasite that invades red blood cells, and transmitted by mosquitoes in tropical regions
9. Parasite that makes a home in the intestine
10. Living things
11. Parasitic insect which feed on the blood of animals
16. Social insects that live in colonies

Answer on page 326

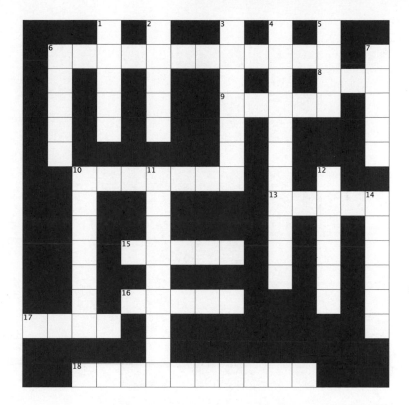

ACROSS

6. Practice of garden maintenance
8. Period, era
9. H_2O
10. Digging tools
13. Arboreal plants
15. Some medicinal plants
16. Labyrinths
17. Mark, indication
18. Glass building where plants grow all year round

DOWN

1. Lawn
2. Evergreen coniferous trees
3. Roses, tulips, and orchids, for example
4. Gardening
5. "A Partridge in a _____ Tree"
6. Gardener's aid
7. Wild-growing plants that are pulled from a garden
10. Sowing
11. Carrots or broccoli, for example
12. Boundaries formed by bushes or shrubs
14. Woody plants smaller than trees

Answer on page 326

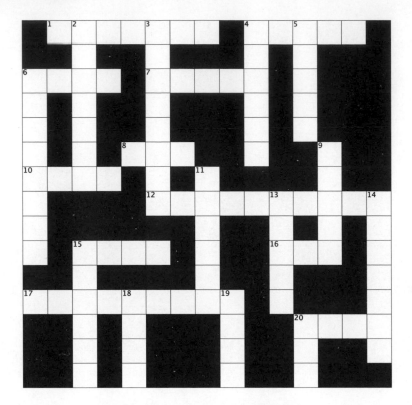

ACROSS

1. Ants, for example
4. Creeping animals with long slimy bodies
6. Type of ants named for their aggressive foraging, known as "raids" a huge number
7. Work
8. Hole in the ground
10. Agreeable
12. Animals that feed on carrion; foragers
15. Type of ants with painful and sometimes dangerous sting
16. It provides light and warmth to the Earth
17. Species of ant that burrows into wood to nest
20. Region

DOWN

2. Wandering
3. Units around which ants organize their lives; settlements
4. Type of ant that forage for food and defend the colony
5. What army ants go on, like soldiers
6. Sensory appendages on the heads of insects
9. Ruler of the ant colony
11. Just-born ants, from the eggs laid by the queen
13. Small ant mounds or bird homes
14. Category below genus
15. Search for food
18. Takes in food
19. Wander
20. Place where animals are kept

Answer on page 326

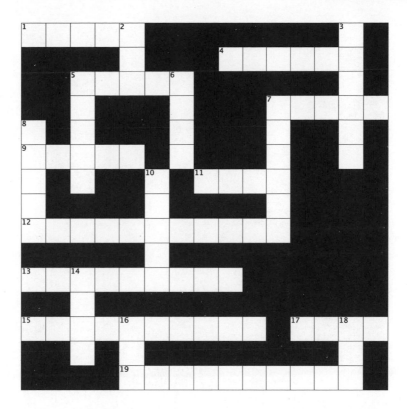

ACROSS

1. Social winged insect that will sting if provoked
4. Type of bee assigned the role of foraging for pollen and bringing it back to the hive
5. What bees make
7. Injury from a bee
9. Ruler of a beehive
11. Sex of the drone bee
12. Job done by bees
13. Hive worker
15. Branch of zoology concerned with insects
17. Epochs
19. The science of poisons

DOWN

2. Foraging bees use this celestial body as a reference point
3. Male bees whose role is to fertilize a queen
5. Homes for bees
6. Lawn
8. Arm
10. Bee movement to indicate to the hive where a food source is located
14. Takes in food
16. Small rug
18. Whichever

Answer on page 326

ACROSS

1. Lions, tigers, and bears, for example
4. Canines
7. A cat or lion, for example
8. Relating to cows
9. A dog, for example
12. Domestic fowl
14. Characteristic of a faraway place
15. Pigs
16. Doctor (including veterinary) who operates
17. Steers
19. Lizard or snake, for example
20. They live wholly in water

DOWN

2. Bird-like
3. Perceive
5. Male goose
6. Fictitious doctor who can speak to animals
10. Relating to horses
11. What doctors' practice
14. Road
15. Reptiles such as rattlers
18. Garlic relative

Answer on page 326

ACROSS

3. The joints between the thigh and the lower leg

7. Tendons at the back of a knee, commonly pulled in sports

8. Limbs for walking

9. Body part susceptible to a boxing injury

10. The grasping organ, often injured in baseball

12. The two joints between the forearm and the upper arm

17. Opposite of far

18. Hang down loosely

21. Protruding part of the jaw

22. Broken bones

DOWN

1. It is worn on the face in fencing

2. Temporary unconsciousness caused by a strike or blow to the head

4. It connects the head and body

5. Upper joints of the arms; often dislocated in some sports

6. Area of the hip between the stomach and thigh that can be injured in many sports

11. Contusion, hematoma

13. They are fractured or broken often in sports like football or hockey

14. Injuries to a limb or muscle by twisting it awkwardly

15. These result from twisting an ankle or wrist

16. Digits

19. Opposite of front

20. Mark left on the skin from a wound

Answer on page 327

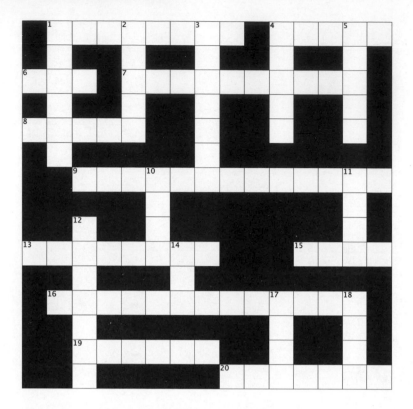

ACROSS

1. Continuance of existence
4. Total
6. Seed, spread
7. Introduce semen
8. Unborn offspring
9. Reproduction
13. Flat organ in which nutrients pass from the mother to the fetus
15. Uterus
16. Animal possessing both male and female sex organs
19. Fetus, fertilized egg
20. Period during which children reach sexual maturity

DOWN

1. Single-celled reproductive units, characteristic of some plants, fungi, and protozoans
2. Streaks
3. Like reproduction for some organisms
4. Organ that produces gametes, such as a testis or ovary
5. Male reproductive cell
10. Pubescent child, adolescent
11. Female reproductive cell
12. Sex cells
14. Knock
17. Dull, plain
18. Release

Answer on page 327

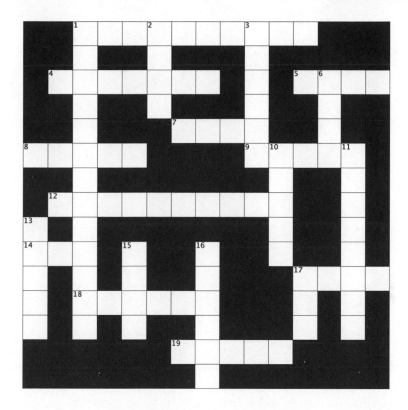

ACROSS
1. The process of burning something
4. Large fire (Italian term); the first part of Dante's *Divine Comedy*
5. Wood structure that is part of a funeral rite
7. What fire provides
8. Fiercely burning fire; conflagration
9. Our planet
12. Greek god expelled from Mount Olympus, for bringing fire to humanity
14. Crater
17. Extremely dry
18. Gas that supports the chemical processes that occur during fire
19. It is generated by fire; blaze, conflagration

DOWN
1. Blaze, inferno
2. Scald
3. Set a fire, kindle
6. Tent-like dwelling
10. Crime of intentionally setting fire to property
11. Atmospheric moisture
13. Flash
15. They are given off by the Sun
16. Wicked light source
17. Everything

Answer on page 327

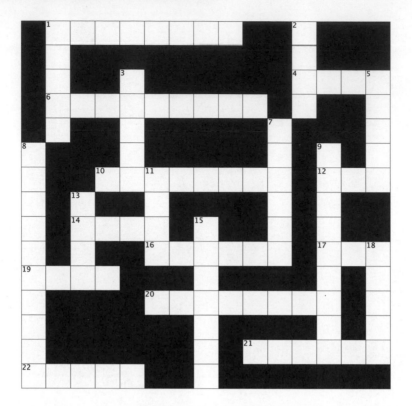

ACROSS

1. Narrow fissures, especially in a rock or wall
4. Beauty queen's accessory
6. Pine tree, for example
10. Dense bushes or trees, in which some animals can take refuge
12. United
14. Unit of land
16. Termite homes
17. Single-story building of crude construction
19. Instance
20. Continuity of existence
21. Difficult situation
23. Quiet corners

DOWN

1. Chambers in a hillside or cliff
that can provide shelter
2. Bird's shelter
3. Land covered with undergrowth, small trees, and shrubs
5. Dwelling
7. Shrubs with stems of moderate length
8. Sanctuary
9. Mounds of loose soil raised by small burrowing mammals
11. Object
13. Beavers' creation
15. Holes dug by some animals, as dwellings
18. Arboreal plants, which provide shelter to animals such as squirrels

Answer on page 327

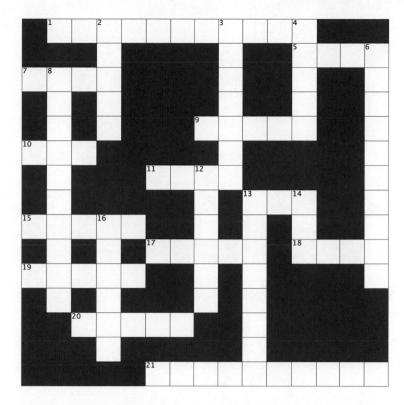

ACROSS

1. Science concerned with designing machines and structures
5. Semi-circles, for instance
7. Unorganized facts
9. Remedies
10. Tooth of a gear or the gear itself
11. Automobiles
13. Average on a course
15. The basic units of matter
17. Implements carried by hand
18. Create
19. Pins that screw into a nut; used to fasten things together
20. Locomotive
21. Branch of knowledge concerned with engineering or applied science

DOWN

2. Small toothed wheels that alter the speed of a vehicle engine and the speed of the wheels
3. Restorations
4. Forms of structured play, such as sports or chess
6. Buildings
8. System of manufacturing that reduces human intervention
12. Machines capable of carrying out actions automatically; often designed with a humanlike form
13. Location, locus
15. Particles
16. Machines that drive automobiles

Answer on page 327

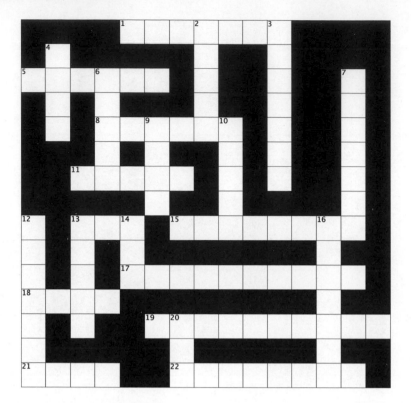

ACROSS

1. Missiles, projectiles
5. Physical harm
8. Weapons used in a biathlon
11. Weapon with a long metal blade; used in fencing
13. Baseball need
15. Guns, rifles, etc.
17. Study of firearms and the effects of bullets or cartridges
18. Opposite of gain
20. Bullets and shells
22. Runner's competition
22. Small bombs thrown by hand

DOWN

2. Weapon also found in the kitchen
3. Long-barreled firearm
4. Injury
6. Ammunition for a bow
7. Gun projectiles
10. Pole weapon consisting of a shaft with a pointed head; used regularly by ancient armies
12. Type of weapons based on atomic energy
13. Explosion, such as that that is caused by a bomb
14. Container
16. Type of gun that fires bullets in rapid succession
20. Russian fighter aircraft

Answer on page 327

ACROSS

5. The study of air motion as affected by a solid object
7. Airliners that do not use propellers
8. They keep a plane in the air
10. Delay
11. Pointer
16. Violent or unsteady movement of air during flight
18. A primary color
19. The main body of an airplane
20. Where the pilot is

DOWN

1. Voyage
2. Opposite of take-off
3. Devices used to reduce the stalling speed of an aircraft wing
4. Passenger plane
5. Force that opposes the aircraft's motion
6. Speed greater than that of sound
9. Accumulate
12. Device used to steer an airliner
13. Rocket propulsion
14. Gasoline, for example
15. Opposite of back
17. Decree

Answer on page 328

ACROSS

3. Aquatic vessel
6. Freighters
7. Marine
9. Small water vessel
10. Night sky bodies, used as reference points in navigation
11. Steep
12. Opposite of behind
13. Device that shows the directions
15. Rain hard
17. It specifies the east–west position of a point on the Earth's surface
18. Pole used to row or steer a boat
19. Recognizable feature used for navigation
20. Drag

DOWN

1. Steer
2. Opposite of near
4. Open boats propelled by paddles
5. Boat used for commercial fishing
6. Distress signal
8. It specifies the north–south position of a point on the Earth's surface
11. Arcs
14. A rebellion of sailors against their officers
15. Person who attacks ships at sea
16. Drench

Answer on page 328

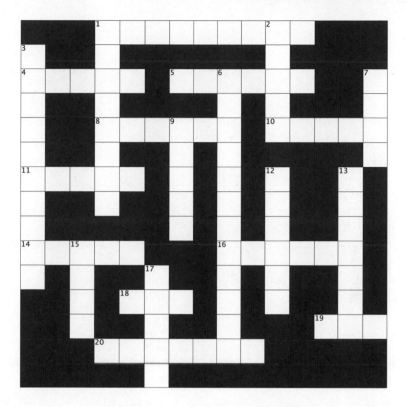

ACROSS

1. Architect's design or plan
4. Space just below the roof
5. Colonnade
8. Plan
10. Steeple
11. Strong string
14. Classical order of architecture
16. Pillar
18. Top
19. Web
20. Veranda

DOWN

1. Edifice
2. Knots
3. Railing used as a parapet on a balcony, bridge, or terrace
6. Structure
7. Space
9. Classical order of architecture
12. Column
13. Molding around the wall of a room just below the ceiling
15. Structure covering a building
16. Structure that encloses the end of a pitched roof

Answer on page 328

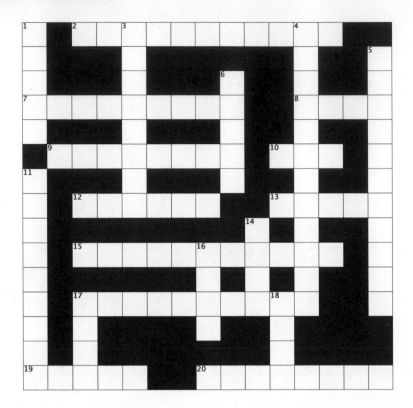

ACROSS

2. Picture writing
7. Wedge-shaped writing used in Mesopotamia, impressed mainly on clay tablets
8. Drops from the sky
9. Disorder in reading words correctly
10. Era
12. Confidential
13. Forms
15. Stylish, decorative handwriting
17. Writing that uses letters representing speech sounds
19. Compete in a bee
20. Ability to read and write

DOWN

1. Typical color for ink
3. The alphabet used by Slavic peoples
4. Writing used in Ancient Egypt
5. The science of language
6. Smudge
11. The study of speech sounds
14. Before the present
16. Talks
17. Adept
18. Doing nothing

Answer on page 328

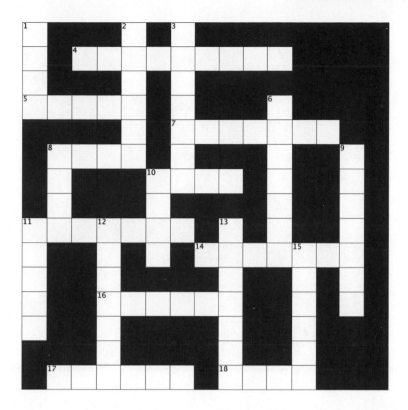

ACROSS

4. Violent wind storm
5. "____ wind," wind blowing steadily from the northeast or the southeast to the equator, especially at sea
7. Wind of the Rocky Mountains, named for a Native American tribe
8. Forceful wind
11. Test yourself (*abbr.*)
10. Strong, cold wind blowing in the upper Adriatic
11. Stopping place
14. Cyclone
16. Sudden wind gust of wind or storm, bringing rain or snow
17. Whirlwind
18. Courage and resolve

DOWN

1. Type of storm caused by wind blowing in a dry region
2. A gentle wind
3. Hot wind that blows from North Africa across the Mediterranean to southern Europe
6. Seasonal prevailing wind in Southeast Asia
8. Burst of wind
9. Tropical storm in the western Pacific ocean
10. Gust
11. Vibrate
12. Prairie storm
13. Sounding like a wolf

Answer on page 328

ACROSS

2. To combat pollution, scientists are looking for ways to harness the energy of this star
8. Pollution
10. Bother
11. Garbage
15. General weather conditions
16. Opposite of dirty
17. Converting waste materials into new ones
19. Distress signal
20. Pollution-causing discharges

DOWN

1. "_____ pollution," harmful levels of sounds
3. Vessel
4. Earth
5. Like uranium
6. H_2O
7. Airplane route
8. Opposite of sea
12. Air pollution
13. Power
14. It blows
15. Decayed organic material
17. "_____ against time"
18. Automobiles

Answer on page 328

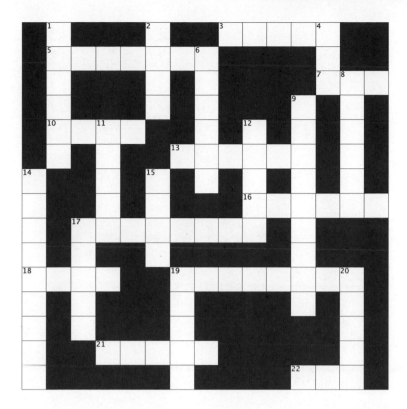

ACROSS
3. Bit of snow
5. Fluids
7. "____ Maria"
10. Barriers constructed to hold back water, forming reservoirs
13. Features of Venice
16. Whirlpool
17. Barometer measurement
18. Epithet
19. System of pipes, tanks, and fittings required for the water supply in a building
21. H_2O
22. Lubrication

DOWN
1. Liquids
2. Rising and falling of the sea
4. Org. concerned with air quality
6. Small river
8. Capacity
9. Source of water
11. Opposite of minor
12. Water regulator
14. Firing on all ____
15. Reddish-brown corrosion caused by water
17. Devices using suction or pressure to raise or move water
19. Tubes used to move water
20. Barbecue

Answer on page 329

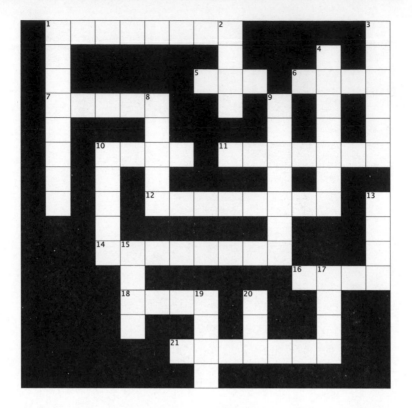

ACROSS

1. Ascending
5. Convene
6. Anchored float
7. Prehistoric invention
10. Appendages used for walking
11. Moving faster than walking
12. Jogging, like a horse
14. Propelling oneself through water
16. Tag, for example
18. Page
21. Bounding

DOWN

1. Moving on all fours
2. Manner of walking
3. Traveling through the air
4. Vaulting
8. Opposite of dark
9. Plunging forward
10. Members
13. Wander
15. Promenade
17. Very eager
19. Places for shoes
20. Move like a rabbit

Answer on page 329

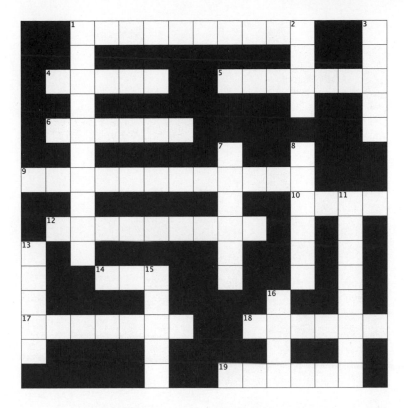

ACROSS

1. The science of the mind
4. Cerebral organ
5. Visualize
6. Think logically
9. The "I" in "IQ"
10. Kind
12. Thought in general
14. Mental acuity
17. Partial or total loss of memory
18. Make up one's mind
19. Deliberate

DOWN

1. Discernment
2. Hindu spiritual discipline
3. Sigmund, who is considered the "father of modern psychology"
7. Recollection
8. A person with exceptional mental ability
11. Think carefully
13. Thoughts, mental impressions or images
15. To use the mind
16. Swiss psychologist Piaget known for his studies of childhood development

Answer on page 329

ACROSS

5. French philosopher and mathematician René, who once proclaimed "Cogito, ergo sum" (I think, therefore I am)
6. Prejudice
8. Objective
9. Nervousness, unease
12. Mend
13. Becoming older
15. Opposite of no
17. "Oedipus ____," Freud's theory that children develop differential relations with each parent according to their sex
18. Establishment where patients can go to seek psychological advice
19. They come to us during sleep
20. Guarantee

DOWN

1. Comportment
2. Sigmund, who was the founder of psychoanalysis
3. Memory
4. Character of a specific individual
6. Opposite of detriment
7. Outlook
10. Natural tendencies
11. Device for weighing things
14. Plenty
16. Apiece

Answer on page 329

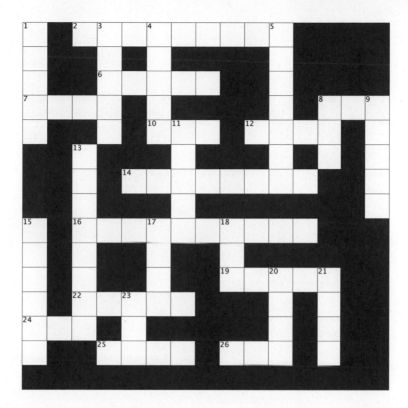

ACROSS

2. Joyous feeling
6. Deep sorrow
7. Opposite of hate
8. Creative activity
10. Female sheep
12. Primal emotion induced by danger or threat
14. Envy
16. The science of the mind
19. It plays out in our mind during sleep
22. Humiliation
24. Strong anger
25. Sense
26. Timid

DOWN

1. Feeling of culpability
3. Rage
4. Conceit
5. Melancholy
8. Exist
9. Have faith in someone
11. Fury
13. Feeling of astonishment
15. Charles, who was famous for his theory of evolution, was one of the first to study the emotions biologically
17. Tranquil
18. Top
20. Green-eyed monster
21. Disposition
23. Wonder

Answer on page 329

ACROSS

1. Fruit that has many symbolic meanings, including primordial knowledge
4. "Carbon ____," Technique used to determine the age of organic matter
7. Archeological digs
10. Place where there are physical remains of past human artifacts and activities
11. Belonging to the distant past
13. Stones, pebbles
15. Study of the past
17. Artifacts
19. Glass containers
20. Relics, remnants
21. Competitive activities, including sports

DOWN

2. Period before written records
3. Age
5. Objects with historical value
6. Image of a divinity as an object of worship
8. Decorative containers
9. The customs, arts, language, etc. of a specific group of people
12. Craniums
14. Pictography and hieroglyphics, for example
16. Natural early human dwellings
18. Water containers

Answer on page 329

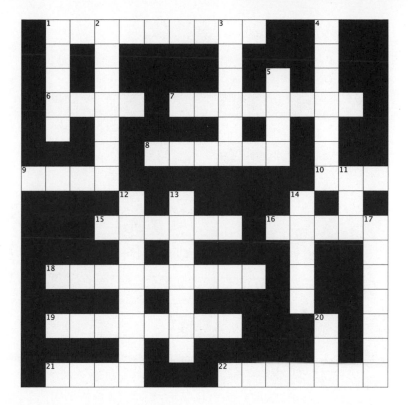

ACROSS

1. Groups of people living together with the same laws and customs
6. Music, painting, and dance, for example
7. It is spoken and written
8. Locations
9. Solemn ceremony
10. Collection
15. Parents and children
16. Opposite of chaos
18. Period of romantic relationship prior to marriage
19. All people
21. Epithet
22. Signs, emblems

DOWN

1. Heroic tales
2. What anthropologists primarily study
3. Cultural background
4. Beginnings
5. Signals
11. Night before
12. Union of two people
13. Blood relationships
14. Closely related group
17. Regularly-occurring ceremonies in a specific culture
20. Before, in the past

Answer on page 330

ACROSS

1. Word with solar or storm
2. What people play
6. Yardstick, benchmark
7. Process of schooling
9. Draws out
10. Spiritual faith
12. Make connections
15. Clay brick material
16. Social standing
17. Country

DOWN

1. Civilizations
3. Study of the production, consumption, and conveyance of wealth
4. Means of communication
6. Customs and conventions of a society
8. Affairs of state, government
11. From the city
13. Work
13. Diversity
14. Aggregate

Answer on page 330

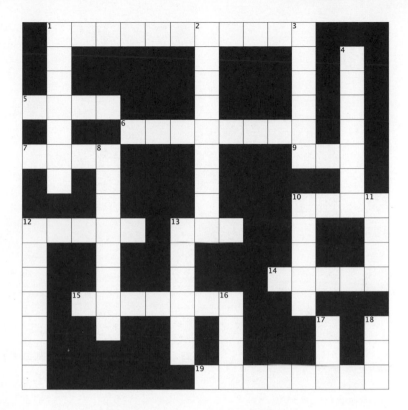

ACROSS

1. The science of language
5. They may be locked or licked
6. Representing speech sounds
7. Tenor
9. That man
10. Go, run, or talk, for example
12. Put forward an idea
13. Inquire
14. Past or present, in English class
15. Local version of a language
19. Like some schools or airports

DOWN

1. Wordlist, dictionary
2. The study of linguistic meaning
3. Discourse
4. Set of rules for constructing words and sentences
7. The phonetic category "dental" refers to *these*
8. Omission of words
10. I, for one
11. Root or stem of a word
12. You or me, for example
13. Adjective modifier
16. It is said to heal all wounds
17. Able to
18. Everything

Answer on page 330

ACROSS

1. Expressive movement
5. Beam
7. Indicate, using the index finger
9. Like Mona Lisa
11. Limb used in gesturing
12. They move instinctively as we speak
14. Curtsy
16. It grows on the body
21. Stance
22. Move into a sloping position
23. It is used to nod, for example

DOWN

2. Signal
3. They are used not only to see, but to express emotions
4. Digits
6. Like some communication
8. Make contact with the hands
10. Not automatic
13. Seal a deal
15. Crying
17. Sit for a painting
18. Grasp
19. Colleague
20. Pedestrian's signal

Answer on page 330

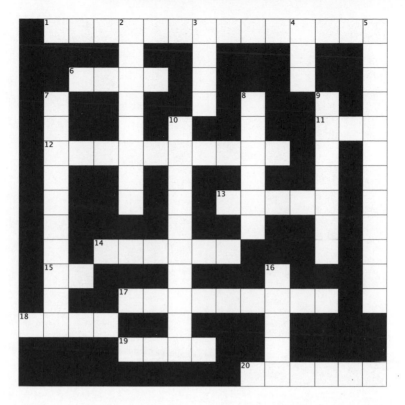

ACROSS

1. Therapeutic discipline Freud founded
6. Cradle
11. Freud's term for the self
12. Science of the mind
13. Assemblage of people
14. They unfold during sleep, which Freud believed were indicators of emotional states
15. Freud's term for the part of the mind where the instinctive impulses are located
17. Period of development that Freud saw as formative of later emotional life
18. Once again
19. Swiss psychologist Carl, who is often associated with Freud
20. Irrational fear

DOWN

2. _____ psychology, branch concerned with the treatment of mental illness
3. Freud's daughter, who continued her father's work
4. Personal pronoun
5. Term Freud used to refer to the part of the mind of which we are not aware but which influences our behavior and feelings
7. Act of excluding desires from consciousness
8. Freud famously smoked these
9. A childhood complex that Freud named after a Greek legendary king
10. Driving impulse to do something
16. Piece of furniture Freud used as part of psychoanalysis

Answer on page 330

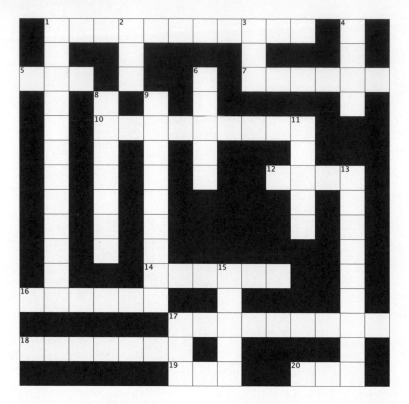

ACROSS

1. There is one in each brain hemisphere, thought to control emotions and memory
5. Number of years someone has lived
7. Psychological wound
10. Techniques for improving and assisting memory
12. "_____-term memory," memory that involves an extended period of time
14. Memory
16. Exit
17. Memoir
18. Class that includes humans and apes
19. Epoch
20. Nudge, as someone's memory

DOWN

1. Underlining
2. Writer Edgar Allan, who often treated aspects of memory in his works
3. Place
4. As originally defined by Richard Dawkins, this refers to a "unit of cultural memory;" now, used mainly to describe an Internet phenomenon
6. "_____-term memory," temporary recall
8. Partial or total loss of memory
9. Recollects
11. Stock
13. February forecaster
15. Scent
17. Insect with a sophisticated memory system

Answer on page 330

ACROSS

3. The "Q" in "IQ"
6. Encephalon
8. Brainiac
9. Oracle
10. Intuition, discernment
12. Debates
13. Acquire knowledge
14. Accomplish
16. Mental sharpness
19. Imaginative

DOWN

1. Puzzles, enigmas (requiring solutions)
2. Fixed set of attitudes or assumptions
4. Comprehend
5. Using the mind
7. Possessing intellectual brilliance
11. Gifted
13. "_____ thinking," often called "thinking outside the box"
15. Clever, intelligent
17. Intelligent
19. Shrewdness

Answer on page 331

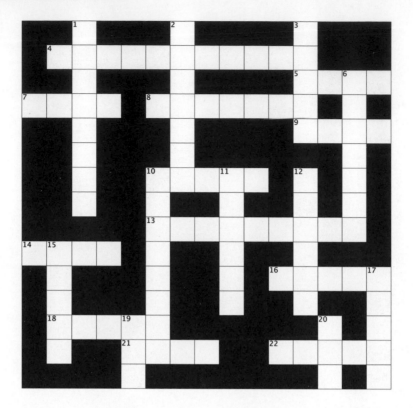

ACROSS

4. Growth, maturation
5. Creative activities such as music and painting
7. Pursue
8. Places for education
9. Perceive
10. Apply oneself in learning something
13. University instructor
14. Increase
16. Where college students might reside on the campus
18. Systematic reasoning
21. Show literacy
22. Seventh sign of the zodiac

DOWN

1. Recall
2. Idea, notion within a certain discipline
3. Hide away
6. School instructor
10. Intuitiveness
11. Describe precisely
13. Knowledge
15. Fruit that is a symbol of knowledge
17. Tussle
19. Wrath
20. Watson's creator

Answer on page 331

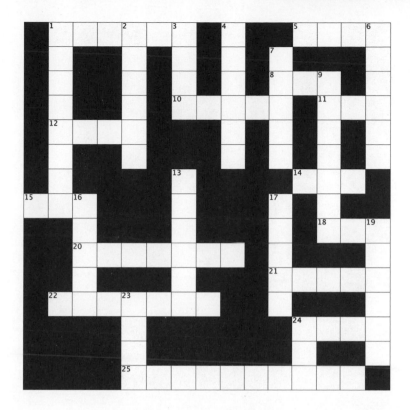

ACROSS

1. Referring to the mouth
5. If the tongue turns this color, it could be a sign that there is insufficient oxygen in the body's tissues
8. Consume
10. The most famous one is seen in da Vinci's *Mona Lisa*
11. Appendage
12. Release
14. Drink small amounts
15. Mandible
18. Gingiva
20. Sharp tooth at the front of the mouth
21. Reef material
22. Tooth doctor
24. Smell
25. Inflammation of the gums

DOWN

1. Bacilli, pathogens
2. Opening
3. Kissers
4. Spit
6. Glossy substance that covers the crown of the teeth
7. They allow us to chew food
9. Speaking
12. Tokens of affection
16. Optimal tooth color
17. Hit
19. Grinding teeth at the back of the mouth
23. Zesty taste
24. Opposite of in

Answer on page 331

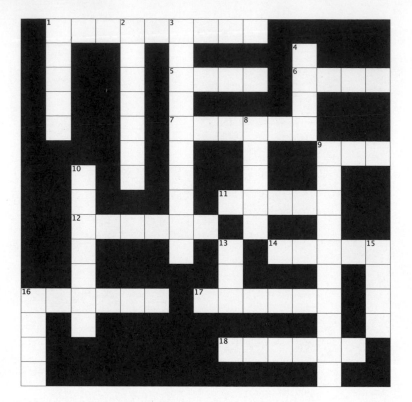

ACROSS

1. Profession with spectacles
5. Looks at
6. Ring-shaped membrane behind the cornea
7. They can correct vision
9. Number of lived years
11. Vision test tool that has letters of different sizes
12. Layer forming the front of the eye
14. Opposite of dark
16. Motionlessness
17. Layer at the back of the eyeball where images are formed
18. Nearsightedness

DOWN

1. Nerve that transmits impulses from the retina to the brain
2. Eye doctor
3. Corrective lenses in a frame
4. Eyeball covers
8. Vision
9. Defect in the curvature of the eye, which results in distorted images
10. "_____ degeneration," degenerative condition affecting the central part of the retina, which might result in vision loss
13. Perceive
15. Physical expression of sadness or happiness
16. Clairvoyant

Answer on page 331

ACROSS

1. Breathing organs
3. Extent
6. Before the present
7. Term
8. _____ of the tongue
9. Roof of the mouth
13. Articulate
14. Creative activity
16. Informal conversation
17. Pitch pattern
20. Art of formal speaking
21. Big
22. Range of a voice or instrument

DOWN

1. Backtalk
2. Stuffy-sounding
4. Speech sound like b, t, or k
5. Speech sound like a, e, or i
10. Enunciate
11. Choppers
12. Speak
15. Organ needed for speech
18. Growl
19. Informal speech
20. Like some exams

Answer on page 331

ACROSS
1. Scam
6. Crime of setting fire to property
7. _____ shot, photo of a criminal form police records
8. Assessment of a perpetrator's character and personality
10. Forensic _____, the study of the criminal mind
12. Age
14. G-man's org.
15. Plays games of chance
17. Convict, criminal
18. Place where justice is conducted
19. Deranged criminal

DOWN
2. Burglary
3. The "D" in D.E.A.
4. Murder
5. Abduction
8. Study of prison management
9. Type of evidence
11. Behavioral science
13. Attack
16. Attorney
17. Turn over

Answer on page 331

ACROSS

4. Name of the process butterflies go through
6. Sample
9. Ooze
10. Soar
12. Young butterflies
13. Blaze, which attracts moths at night
15. Scarce
18. Semi-circle
20. Leaping amphibians
21. Appendages attached to the head of butterflies used for smelling

DOWN

1. The larva stage of the butterfly
2. Covering that protects the pupa
3. Colorful appendages for butterflies
4. Wool eaters
5. Chrysalis
7. Everything
8. The scientific study of insects
11. Large orange and black butterfly found mainly in North America
14. Scaly-skinned reptiles with moveable eyelids; they prey on butterflies
16. Color between black and white
17. Stay
19. Butterflies need its light and heat to survive

Answer on page 332

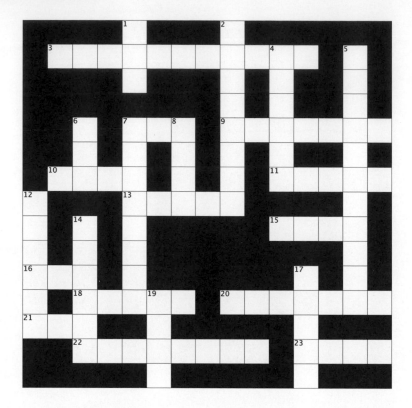

ACROSS
3. Not sleeping
7. When the sun is out
9. Tissues that provide strength
10. Sleep lightly
11. Unaccompanied
13. Temporary cessation of breathing during sleep
15. Repose
16. Consume
18. When the moon is out
20. Sleep sound
21. Fairytale sleep disruptor
22. Prediction
23. Place for a scarf

DOWN
1. Phase of sleep
2. Sleeplessness
4. Midday snooze
5. Winter sleep
6. Exclamation said to scare
7. Fantasizing
8. Involuntary act of inhaling deeply, with open mouth, due to tiredness
12. Opposite of awake
14. Snooze
17. Beverage
19. Optimistic feeling

Answer on page 332

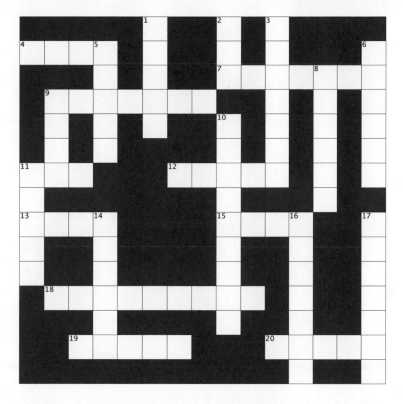

ACROSS

4. Rounded vault that is the roof for some observatories
7. Eighth planet, named after the Roman god of the sea
9. Planet nearest the sun
11. Airport grp.
12. System of planets and their moons in orbit around the Sun
13. Zone area
15. The sun, for example
18. Astrological char
19. Small planetary body that orbits the sun, named for the god of the underworld
20. Word after outer

DOWN

1. The hottest planet, named for the Roman goddess of love
2. Star that is orbited by the planets in this puzzle
3. Largest planet in the solar system, named for the chief Roman god
5. Our planet
6. Cloud of gas and dust in outer space, visible in the night sky as a bright patch or dark silhouette
8. Seventh planet from the sun, named after the first ruler of the universe in Greek mythology
9. The red planet, named for the Roman god of war
10. Instrument which allows astronomers to look into the distant universe
11. What a falling meteor leaves
14. Starry
16. Transform
17. Italian Renaissance scientist who made many discoveries about the solar system

Answer on page 332

ACROSS

6. Tiny materials used to make integrated circuits

8. Subatomic particles that can be either positively or negatively charged

9. Representative image on a computer screen

11. Android designed to assist people

12. Units of digital information

16. Electrical lines

18. Embrace, like new technologies

19. Raw information

20. Sections

21. Device that stops communications

DOWN

1. Motherboard array

2. Semiconductor devices capable of amplification

3. Connected series of metal links

4. Look at

5. Electronic components

7. ____ Valley, area in California famous for its computing and electronics industries

10. TV, slangly

13. Place to surf

14. Empty space

15. The "R" of N.P.R.

16. Electrical units for the rate at which energy is generated

17. Electrical impulse or wave transmitted or received

Answer on page 332

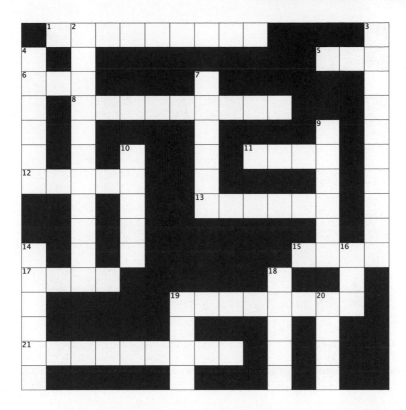

ACROSS
1. Living
5. Lungful
6. A state of matter
8. Of the lungs
11. Weary sound
12. Piece of land
13. Breathe out
15. Predicament
17. Chest organ
19. Trachea
21. Muscular partition between thorax and abdomen

DOWN
2. Breathing
3. Lung inflammation that causes coughing
4. Shocked
7. Breathe in
9. Breathe with a whistling sound
10. Motionless
14. French physiologist Bernard, who important discoveries about the body
16. Visualize
18. Cessation of breathing during sleep
19. Opposite of strong
20. Short, abrupt burst of breath

Answer on page 332

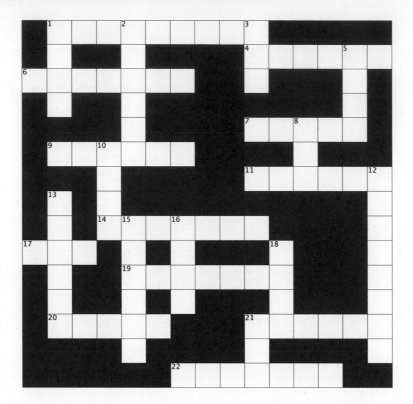

ACROSS

1. Polar region around the South Pole
4. Spain's continent
6. Varied
7. Earth's outermost layer of rock
9. Great White North
11. Tectonic _____
14. North, South or Central
17. Diagram of a geographic area
19. Its capital is Mogadishu
20. Chief city of Nigeria
21. Mali's desert
22. Area encompassing the islands of the Pacific Ocean

DOWN

1. The largest continent
2. The second largest continent
3. Corporate V.I.P.
5. City with a harbor
8. U.N. member
10. Mission control org.
12. China's most populous city
13. Largest country in South America
15. Russian capital
16. Italian capital
18. Spanish house
21. Resort with a mineral spring

Answer on page 332

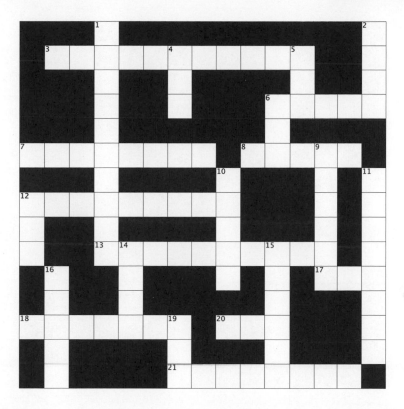

ACROSS

3. The science of maps
6. Collection of maps
7. Cylindrical map projection devised by a Flemish geographer in 1569 and named after him
8. Ratio of a distance on the map to the corresponding distance on the ground
12. Kind of map that shows territorial borders
13. Passage of ships
17. Directly
18. Outline of a natural feature
20. Perceive
22. Imaginary semicircle on the Earth's surface from the North Pole to the South Pole

DOWN

1. Method for representing the surface of the Earth on a map
2. Tops
4. Navigational aid, for short
5. Still
6. Semi-circle
9. Chart that explains the symbols used on the map
10. Opposite of opaque
11. Recurring things
12. Place
14. In addition
15. Have in mind
16. Spherical representation of the Earth with a map on the surface
19. Ewe's mate
21. Opposite of land

Answer on page 333

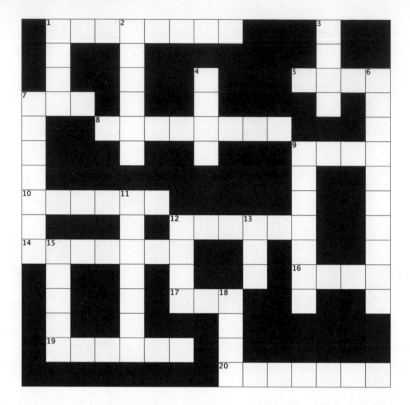

ACROSS

1. Volcanologist's study

4. "Volcano of ____," name given to a Guatemalan volcano

6. When it's light day

7. Active volcano near Naples, which destroyed an ancient Roman city

8. Discharge

9. Like a volcano that is likely to erupt

11. US state with five volcanoes, including Kilauea

13. Destructive, long high sea wave

15. Wine vessel

16. Opposite of high

18. Earth's interior between the crust and the core

19. Ancient city destroyed by a volcano

DOWN

1. Active volcano in eastern Sicily

2. Tectonic ____, which make up the lithosphere

3. Molten rock expelled by a volcano

5. Major tremors

6. Inactive, like a volcano

8. Like a volcano which has not erupted in the past 10,000 years

10. Bridge over a valley

11. Rise

12. Mixture of rock, mineral, and glass particles expelled by an eruption

14. Condensation

17. Distort

Answer on page 333

ACROSS

1. Small narrow river
4. Current
7. Region
9. Rivulet
10. Downpour
12. Proportion
14. Result of mixing of earth and water
15. Small waterway
17. One of two great rivers of Mesopotamia, the other being the Euphrates
20. Spout
21. Deed
22. Principle Chinese river
23. River in southern Asia that flows from Tibetan Region of China through Kashmir and Pakistan to the Arabian Sea

DOWN

2. River flowing into a larger river
3. Imprint
5. Navigable body of water
6. Rome is on its banks
8. Inundate
11. The longest river in the world that was crucial to the foundation of ancient Egypt
13. Torrent
14. Place where a river enters another body of water
16. River flowing through Washington, DC into Chesapeake Bay
18. River in northern India and Bangladesh that is considered sacred to many
19. Preserve

Answer on page 333

ACROSS

2. with Milwaukee and Chicago on its shores
5. Large freshwater lake in the Sierra Nevada
7. Puddle
8. Shelter from the wind
10. Stretch of salt water separated from the sea
11. Narrow river
13. Lake forming a border between France and Switzerland
15. Although called the Caspian _____, it is commonly classified as a lake
16. Largest lake in Africa, lying mainly in Uganda and Tanzania

DOWN

1. Lake that lies between Canada and New York and gets it's name from the province it's in
3. Second-largest lake in North America
4. Third largest of Italian lakes after Lake Garda and Lake Maggiore
6. Make-up of lakes
7. Small body of still water
9. Depression on the Earth's surface, containing water
12. Melting of snow as the weather warms up
13. _____ Lakes - group of lakes in North America
14. The Caspian Sea is the largest lake on this continent

Answer on page 333

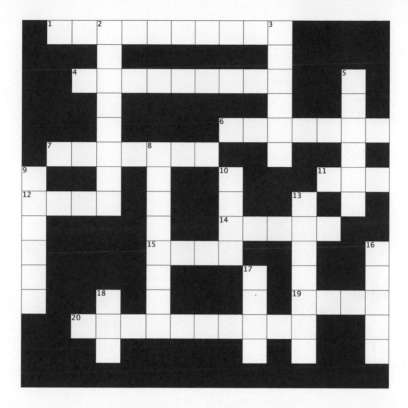

ACROSS

1. The air enveloping the Earth
4. Flash of electrical discharge during a rainstorm
6. Seasonal wind associated with the Indian Ocean
7. Thor's domain
11. The bottom of a river
12. Adoration
14. Disturbance that may have strong winds or rain
15. Come down hard
19. Nothing
20. Like some rainfall

DOWN

2. Condensation
3. Motor
5. Brief, light rainfall
8. Heavy rainfall
9. Sources of rain
10. Haze caused when warmer air meets colder air
13. Fine rain
16. Deposit of ice crystals on the ground
17. Frozen rain
18. Thick cloud of water droplets that reduces visibility

Answer on page 333

ACROSS
1. Radiance
4. Opposite of dark
6. Celestial body that's part of an eclipse
8. Look at
9. Ring-shaped
12. Type of eclipse where the moon appears darkened
13. Go bad
14. Region in Russia and Kazakhstan
16. Eclipse phenomenon
18. Partial eclipses have been recorded for this planet by NASA's Curiosity rover
20. Settlement
21. What eclipses have been thought to be

DOWN
2. Celestial body that's part of an eclipse
3. Lining up
5. The shadow cast during a partial eclipse
6. Eclipse of the sun
7. Type of eclipse that's not full
10. Singular
11. The science of outer space
15. Type of eclipse that occurs when the Sun or moon are completely obscured
17. Atmosphere
19. Total

Answer on page 333

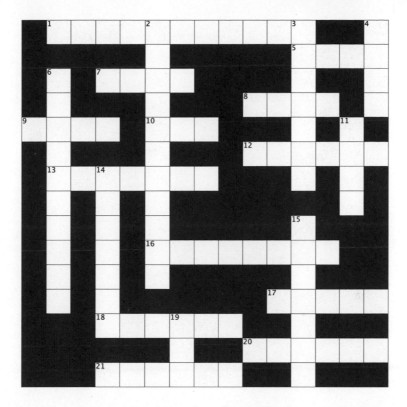

ACROSS

1. Great apes native to the forest and savannah of tropical Africa
5. Very small amount
7. Protruding facial feature unique to humans
8. Manner of walking
9. Prehensile part of the arm
10. General name for nonhuman primates such as gorillas, chimpanzees, and orangutans
12. Small primate with a long prehensile tail
13. Slender, tailless tree-dwelling ape that emits loud hooting calls
16. The largest living primates
17. Implements (which some primates, like chimps, can make and use)
18. Tree-dwelling primates found only in Madagascar; they have a pointed snout, large eyes, and a long tail
20. Pygmy chimps that are native to rainforests of the Democratic Republic of Congo
21. Homo sapiens

DOWN

2. Branch of zoology dealing with primates
3. Apes or monkeys in general
4. It grows on primates
6. Large arboreal apes with long reddish hair and long arms, native to Borneo and Sumatra
11. Limbs for walking and running
14. Two-legged
15. Old World monkeys distinguished by naked callosities on the buttocks
19. _____ Today

Answer on page 334

ACROSS

1. Derived from living matter
8. Toppers
9. The feel or consistency of a substance
10. It is found on beaches, riverbeds and deserts
12. Curriculum component
14. Loose soil
16. Solid inorganic substances
18. Stones and pebbles
20. An organism's natural home or environment
21. Wetland

DOWN

2. Atmospheric mix
3. Acorn, for one
4. Dense
5. Dried stalk of grain
6. Surface of the Earth
7. Soil surface consisting of decomposed organic matter
11. Opposite of far
13. Flora
15. Plowed layer
17. When mixed with soil it forms mud
18. Underground parts of plants
19. Sticky soil which hardens when dry

Answer on page 334

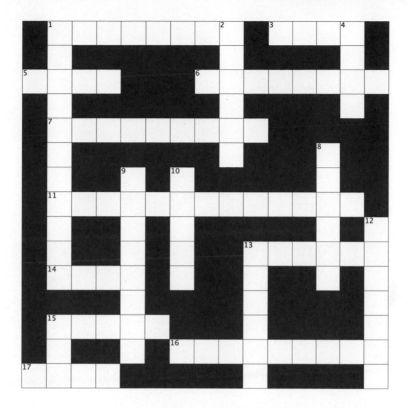

ACROSS

1. Vitamin or mineral, for example
3. Groceries
5. Food with low nutritional value, such as packaged snacks
6. Energy we get from food and drink
7. Intake

11. Kind of diet high in vegetables, fruits, whole grains, and olive oil
13. Substance in cereal grain that causes illness in people with celiac disease
14. Fit

15. Acids that occur naturally in plant and animal tissues
16. Expert on nutrition
17. Carnivore's delight

DOWN

1. Sustenance
2. Home brewer
4. Regime of foods eaten regularly

8. Wellbeing
9. An essential nutrient
10. One who does not eat or use animal products
12. Extremely old
13. Feeling after eating too much dessert, perhaps
15. Number of years lived

Answer on page 334

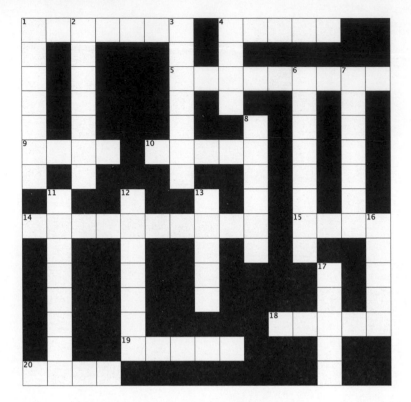

ACROSS
1. Living organisms distinguished from plants
4. ____ & flora
5. Environmentalist's concern
9. Common rodents with pointed snouts
10. Striped African horse
14. Defense
15. Small social insects
18. Large predatory marine fish
19. Type of salamanders, with lungs and a well-developed tail
20. Arboreal plant

DOWN
1. They are found on deer, elk, caribou, and moose
2. Bees, spiders, ants, and mosquitoes, for instance
3. Organisms with a shared evolutionary history
4. Tailless amphibian with long hind legs for leaping
6. Propensity to continue living
7. Animals that no longer exist
8. Wolf or fox
11. Animal that preys on other animals
12. Flightless aquatic bird
13. Large, striped feline
16. Bad-smelling animal
17. Large marine mammal

Answer on page 334

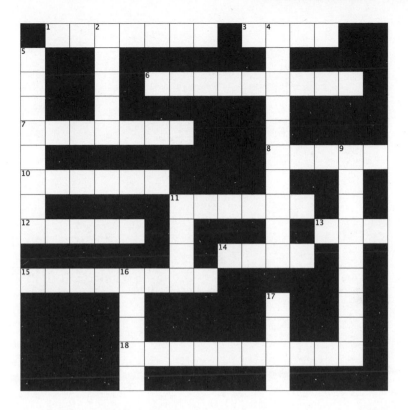

ACROSS

1. Science of animals
3. Nocturnal mammal capable of flight
6. Four-footed animals
7. Tortoises
8. Wrens, robins, and blue jays, for example
10. Animal with long body and tail, movable eyelids, and scaly skin
11. Warm-blooded vertebrate animal
12. Viper or cobra, for example
13. Famously cunning canine
14. Cod, salmon, or herring, for example
15. Spine
18. Predatory semiaquatic animals with long jaws and a horny textured skin

DOWN

2. Nocturnal newborn
4. Frogs, toads, and salamanders, for example
5. Snakes, lizards, and turtles, for example
9. Large extinct reptiles
11. Primary
16. Stabilize
17. Turn over

Answer on page 334

ACROSS

1. Arthropods such as lobsters, crayfish, and prawns
5. Sea creature that moves sideways
6. What entomologists study
9. Tiny arachnids related to the ticks, which live in the soil
10. Household nuisances
11. We breathe it
13. Insect distinguished by hard forewings that cover the hind wings and abdomen
14. Prawns' relatives
15. Help
17. Large insect with long transparent wings, which makes a loud shrill noise by vibrating its abdomen
18. Process of transformation from an immature to an adult form, typical of insects or amphibians

DOWN

1. Common feline
2. Spiders and scorpions, for example
3. Picnic pest
4. Arachnid with pincers and a poisonous sting on its tails
6. An animal lacking a backbone, including arthropods
7. Multi-legged arthropod
8. Ocean
12. It feeds on the skin of dogs and cats
16. Filth
17. Harvest

Answer on page 334

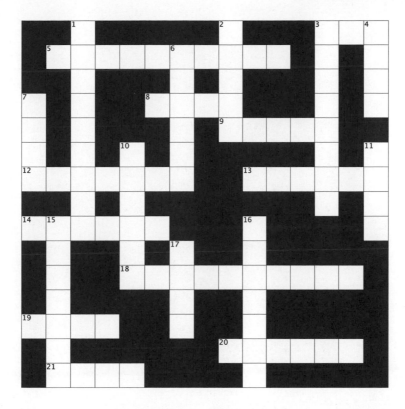

ACROSS

3. Spring month when some flowers bloom

5. Flowers worn symbolically on Mother's Day and wedding anniversaries

8. Color of hydrangeas and irises

9. Bit of holly

12. Plants whose flowers have clusters of yellow petals and white rays

13. Mexican plant of the daisy family, cultivated for its brightly colored flowers

14. Remnant

17. Plants of the daisy family, with large golden-rayed flowers

18. It blows

19. Modified leaves surrounding the reproductive parts of flowers; corolla

20. Above-ground stalk that supports the flower

DOWN

1. Large fragrant white or yellow flowers, native to warm regions

2. Romantic flowers

3. Flower of daisy family, with yellow, orange, or copper-brown petals

4. Twelve months

6. Spring-blooming cup-shaped flowers of the lily family

7. Like uncultivated flowers growing without human intervention

10. Flowers symbolically associated with motherhood, and rebirth

11. Pastel

14. Flowers with lip-shaped petals that are often used in corsages

16. Shrubby plants cultivated for their showy flowers

17. Pansies are hardy winter flowers that can even survive in this winter precipitation

Answer on page 335

ACROSS

1. The study of animal behavior
7. Automatic behaviors
8. Originate (from)
9. Gesture or sound conveying information
14. Message exchange
15. Hindmost part of an animal
17. Gaining someone's love
18. Gregarious
19. Expressive motion
20. Bother

DOWN

2. Opposite of wild
3. Lower extremities
4. Russian physiologist Ivan, who was known for his studies on conditioned reflexes, including his famous experiments with dogs
5. _____-or-flight response
6. Exhibition
10. Rapid learning that establishes behavioral patterns or habits
11. Lizards with the ability to change their color for camouflage
12. Simulation
13. Speak
16. Spontaneous activity in many animals, such as chasing a ball
17. To move the tail to and for

Answer on page 335

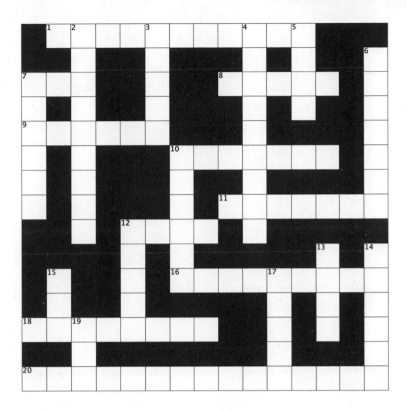

ACROSS

1. Like sparkling stones
7. Back talk
8. A state of matter
9. Hard, white mineral consisting of silicon dioxide, found in many rocks
10. Made of clay hardened by heat
11. Home of Bollywood
12. Precious stone of pure carbon
13. Small talk
17. Feathery ice crystal that falls to the ground in winter
19. Mineral used in jewelry
21. The science of crystals

DOWN

2. Replicating, like the arrangement of atoms in a crystal
3. November birthstone
4. Like minerals
5. Grand
6. Precious green stone
7. A state of matter
10. Weight units for gems
13. Desire
14. Flavorful white crystalline
15. Splendor
16. Number of years a person has lived
18. More elegant
20. Memorial Day month

Answer on page 335

ACROSS

1. Closest to the top of the globe
5. Region around the North Pole
8. Lines that converge at the North Pole
9. Opposite of near
11. A compass needle points to this pole
14. Uncooked
15. Singular
17. Path
18. Large red deer native to North America
19. Thawing
20. Consumes
21. Large floating mass that has become detached from a glacier

DOWN

2. The North Pole is where the Earth's axis of _____ meets its surface
3. Opposite of north
4. Opposite of sea
6. Indigenous people of northern Canada and parts Alaska
7. Explorer Robert, who claimed to have reached the North Pole, on April 6, 1909
8. Opposite of high
10. Legendary inhabitant of the North Pole
12. Clock
13. Explorer Frederick Albert, who claimed to have reached the North Pole on April 21, 1908
14. Sleigh pullers
16. Global _____
17. Large bodies of water surrounded by land

Answer on page 335

ACROSS

1. Opposite of north
6. Repast
7. Crustacean in a cocktail
9. Attractive, as a Pole might be
11. Faster than a walk
12. Course
13. Opposite of south
15. During summer in Antarctica, this never goes down
16. Hours
17. A state of matter
19. Luminous celestial bodies
20. Mount Erebus and Deception Island, for example

DOWN

2. Of the Earth
3. _____ of the iceberg
4. Moves steadily, as a stream
5. Burdensome bird
6. These "falling stars" are common in Antarctica
8. Norwegian explorer Roald who was the first person to reach the South Pole
10. Coming together, as time zones do at the South Pole
12. Black and white flightless seabirds, found almost exclusively in the Southern Hemisphere
14. Fish-eating aquatic mammals with flippers
18. Ross _____ - the arm of the Pacific Ocean that forms an indentation in the coast of Antarctica

Answer on page 335

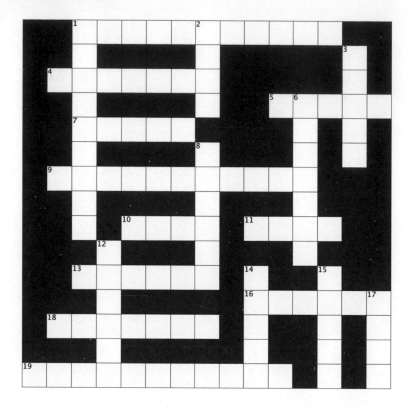

ACROSS

1. Descent through the male line
4. Male sibling
5. A parent's brother
7. Daughter of one's sibling
9. Descent through the female line
10. A parent's sister
11. Group of close-knit and interrelated families, typical in the Scottish Highlands
13. Female parent
16. Son of one's sibling
19. Spouse
20. Parent's parent

DOWN

1. Ancestry
2. Myths or legend
3. Relative by marriage
6. "_____ family," composed of a couple and their dependent children
8. Female sibling
12. Child of one's uncle or aunt
14. Marriage
15. Kid
17. Spouse

Answer on page 335

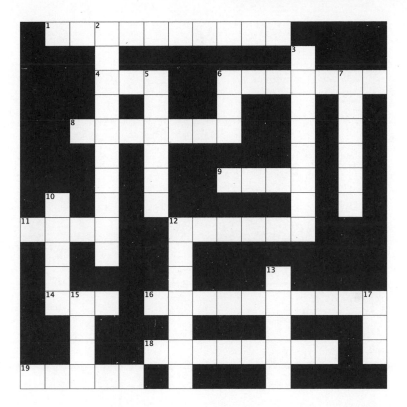

ACROSS

1. Despair
4. Opposite of no
6. Great worry an unease for what might happen
8. Treatment of psychological problems
9. Emotional state
11. Terror
12. Opposite of smaller
14. Acquired
16. Result of isolation
18. Irrational suspicion
19. Sudden intense fear

DOWN

2. Profession treating mental illness
3. Ailment
5. Mental or emotional strain, tension, or pressure
6. Whichever
7. Emotional damage
10. Existing
12. Disorder characterized by alternating periods of joy and depression
13. Not serious
15. Candid
17. Opposite of happy

Answer on page 336

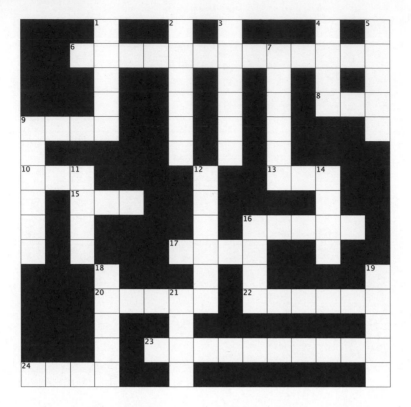

ACROSS

6. The cultivation of trees

8. Ground surface with growing grass

9. Protective outer layer of the tree's trunk, branches, and twigs

10. Silver-gray tree, with green, oval shaped leaves; used commonly to make baseball bats

13. Fluid that comes from trees

15. Tree that gives its name to many a street

16. Tree renowned for its syrupy sap

17. Evergreen coniferous tree with clusters of long needle-shaped leaves

20. They can be seen in the cross section of a tree trunk, representing a single year's growth

22. Tree that is used commonly as a Christmas tree

23. Biodiverse area with a canopy

24. Decorative plant

DOWN

1. The woody stem of a tree

2. Development

3. Smooth oval nuts, from oak trees

4. Sprouts

5. Units of plant reproduction

7. They fall in the fall

9. Arm of a tree

11. Rope fiber

12. Oranges, apples, and pears, for example

14. Evergreen tree with fan-shaped leaves, that grow in warm regions

16. Clutter

18. Color used generally to describe tree bark

19. What birds build on tree branches

21. Color between white and black

Answer on page 336

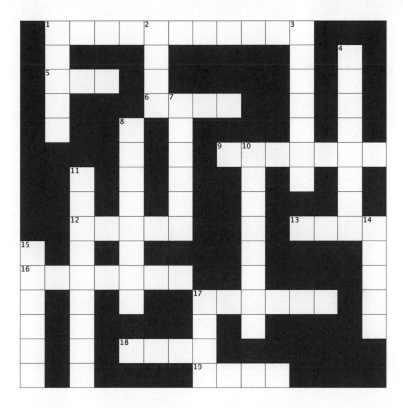

ACROSS

1. Green pigment, present in green plants and algae; essential in photosynthesis
5. Species of algae called rhodophyta that has this color
6. Algae is this for many aquatic animals
9. Sedentary aquatic invertebrate with a porous body for drawing in a current of water from which nutrients are extracted
12. Citrus fruit
13. Climb
16. Living in or near water
17. Element with symbol C
18. Opposite of under
19. Public green area

DOWN

1. Marine formation
2. Great Barrier _____
3. Composite plants emerging from a combination of fungi and algae, which extract nutrients from the atmosphere
4. Natural colors of plants
7. Algae produce this important gas
8. Stick of explosive consisting of nitroglycerine
10. Synthetic organic materials used as plastics and resins
11. Contamination of air, water, soil, etc.
14. Our planet
15. Squanders
17. Freshwater fish, typically with barbels around the mouth

Answer on page 336

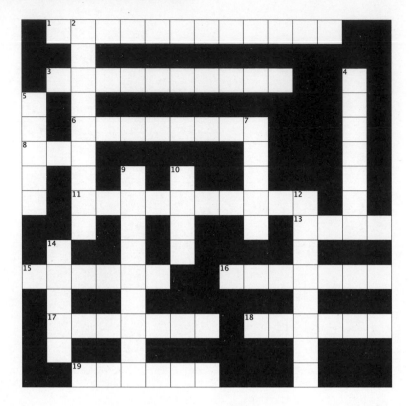

ACROSS

1. Branch of medicine concerned with the incidence, distribution, and control of diseases

3. Transmittable, like a disease

6. Occurrence of a widespread disease in a community

8. Male sheep

11. Communicable

13. Stew ingredient

15. "Bubonic _____," also called the Black Death

16. Bacterial disease contracted from contaminated water

17. "_____ flu," also known as the "1918 flu"

18. Also called "spotted fever," transmitted by lice, ticks, mites, and rat fleas

19. Unchanging

DOWN

2. Diseases that spread throughout the world, including COVID-19

4. Mosquito-borne infectious disease, common in tropical zones, characterized by high fever, tiredness, vomiting, and headaches

5. Pathogen that is the cause of many diseases

7. Disorder

9. Respiratory illness

10. Early 2000s outbreak, for short

12. Variola, pock-mark leaving viral infection, eliminated by vaccination by 1979

14. Opposite of "open"

Answer on page 336

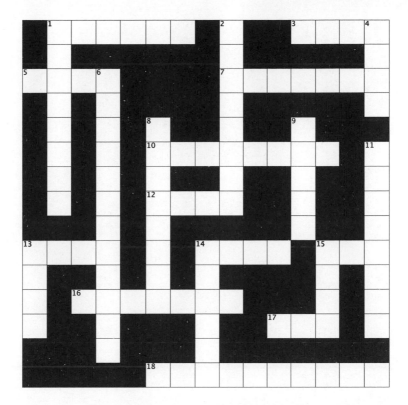

ACROSS

1. Trip by air
3. Pilot's course
5. The world's largest continent
7. Spaceships
10. Pilot's field
12. Elevate
13. Facts and figures
14. One of two on an airplane that support it in air
15. Holder of first-aid supplies
16. Blimp
17. Most common type of airplane
19. Spacecraft pilots

DOWN

1. Main body of an airplane
2. Any machine capable of flight
4. Front part of the airplane
6. Study of the properties of moving air
8. Hot-air crafts
9. Ring-shaped
11. Airplane's height above the ground
13. Force that resists the motion of an aircraft through the air
14. The _____ brothers," invented the first (modern) airplane
15. Unit of speed used in aviation
18. Airport inits.

ACROSS

ACROSS

3. Marine animals with flippers
6. Marine
7. Large marine mammals with a horizontal tail fin, and a blowhole on top
8. Coral ridges above or below the surface of the sea
11. Mollusk with eight tentacles

16. Natural homes
17. Limbless reptiles which may be venomous
19. Pond-dwelling and croaking amphibians
21. Long, fast-swimming mollusk

DOWN

1. Flat shark relative
2. It has gills and fins and lives in water

4. Slippery creatures, some of which are electric
5. Large predatory marine animals
6. Species of marine plankton
9. Mushrooms and molds, for example
10. Highly intelligent whale species, with a beaklike snout and curved fin

12. Collision
13. Aquatic invertebrate that "absorbs" water nutrients
14. Class of water organisms, including bacteria, archaea, algae, protozoa
15. Tailed aquatic larva of an amphibian
18. Turf
20. Opposite of land

Answer on page 336

ACROSS

1. Study of humanity
5. Inherited characteristic
6. Standards
9. Signs that stand for something culturally important
11. Moral and ethical principles
13. Cooking that is characteristic to a certain place
14. Cookware
15. Group with distinct behavior within a larger one
17. Historical culture, passed on from generation to generation

DOWN

1. Creative activity
2. Ceremonies
3. Sonnets and epics, for example
4. Historical culture, passed on from generation to generation
7. Decorative open container
8. Historical tradition, passed down from previous generations
10. Views held to be important by specific groups
12. Point out
13. Habitual practices
16. Formal ceremony

Answer on page 337

ACROSS

1. Acoustics is a branch of this science
3. Loud or unpleasant sound
6. Common term for atmosphere
8. Use of language for purposes of communication
9. Reflected sound
10. High card
12. Listen
13. One who moves to music
14. Vocal inflection
15. Musician's concern
16. Deep, like a musical bass line
17. Unit of frequency
18. Opposite of dirty

DOWN

1. Spelled as spoken
2. Auditory sensations
4. By word of mouth
5. Sound _____, vibrations that generate sounds
7. Quality of sounds that are deep and reverberating
9. Organ with a hammer
10. Able to be heard
11. Amount of time or a particular time interval in music
12. Like a soprano's voice
14. Quality of a musical sound

Answer on page 337

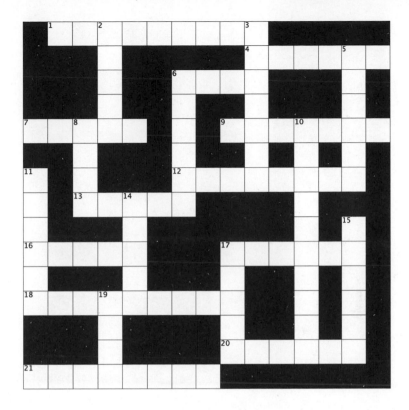

ACROSS

1. Groups of atoms bonded together
4. Contaminated
6. Marine mollusk
7. Strong polymer adhesive
9. Synthetic material made with organic polymers that is used commonly to make products such as food wraps and grocery bags
12. Has seniority
13. Tough, elastic synthetic polymer used in stockings and hosiery
16. Rudimentary
17. Opposite of circular
18. Recurring
20. Camera parts
21. Chains of amino acids that are essential to all living organisms

DOWN

2. Polymer synthetic product used to make paints, gloves, and clothing
3. Alike
5. Viscous substances of plant or synthetic origin that are convertible into polymers
6. Fluffy staple fiber that grows in a boll
8. Kiln
10. Artificial products
11. Tire material
14. Wet suit fiber
15. Vestiges
17. Allowed
19. Opposite of west

Answer on page 337

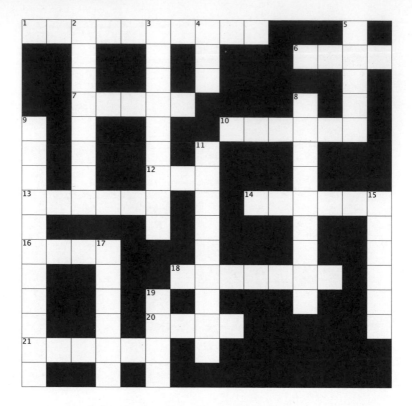

ACROSS

1. Becoming packed together, the way glaciers are formed
6. Falling flakes
7. Courage
10. Long inlets with steep cliffs, created by a glacier and common in Norway and Iceland
12. Melting ice causes these levels to rise
13. Snarl
14. Country north of the US, where many glaciers can be found
16. Agrees nonverbally
18. Mass of rocks and sediment left behind by glaciers
20. Mineral marking
21. Like Swiss mountains, where many glaciers are found

DOWN

2. Everest, for example
3. Deep cracks in glaciers
4. Frozen water
5. The top and bottom of the Earth
8. Large island that lies to the northeast of North America, within the Arctic Circle
9. Type of breakfast
11. Region at the southern end of South America
15. US state on the northwest extremity of the country's west coast home to the largest glacier in North America
17. Narrow passage of water connecting bodies water
19. Flush

Answer on page 337

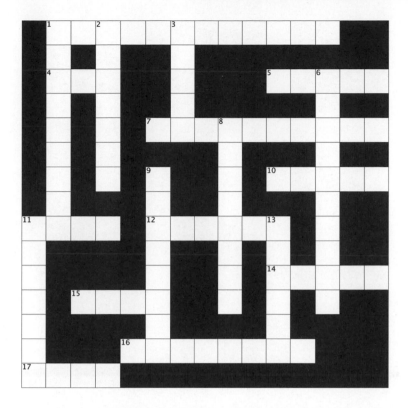

ACROSS

1. Functions
4. Beam
5. Rounded vaults forming the roofs of buildings
7. Powered by current
10. Talons
11. Dip or mark left by a blow to a hard substance
12. Branch of engineering for ships and offshore installations
14. Slow down
15. Discipline that is essential to engineers
16. Difficulties
18. Blueprint or diagram

DOWN

1. Branch of engineering for flying
2. Science of matter and energy
3. Branch of engineering for the construction and design of roads, bridges, dams, and other structures
6. Branch of engineering for the construction and design of machines
8. Branch of engineering for the design and operation of certain plants
9. Branch of engineering for the design and operation of systems and devices
11. Expand
13. Insignia

Answer on page 337

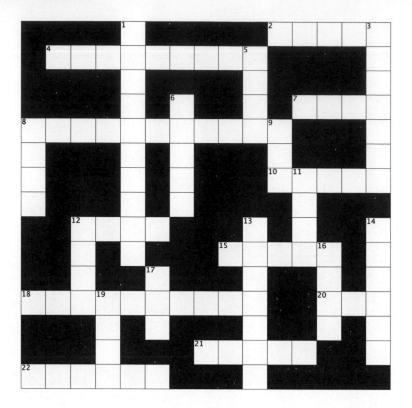

ACROSS

2. Knuckle or knee
4. Palm-reading
7. Name for a finger or what you wear on one
8. They are cut or clipped
10. Hand-forearm connector
12. Fortune-teller's interest
15. Opposable digit
18. Bone of the hand
20. Approval
21. Longer of the two bones in the human forearm
22. Small digit

DOWN

1. Capable of grasping
3. Diverging line
5. Shout
6. Pointer
8. Clenched hand
9. Stitch
11. Space
12. Section
13. Bone of the finger
14. Finger that can offend
16. Wedding _____
17. Strike lightly
19. Wildly

Answer on page 337

ACROSS
1. Message exchanges
5. Roman emperor who used a cryptographic system that shares his name
8. Numbers divisible by only 1 and themselves
9. Creative activity
10. Confidential
13. Exact
14. Collection
17. German military code cracked at Bletchley Park during World War II

DOWN
1. The art or science of deciphering coded messages
2. Discipline that is central to cryptography
3. Things that can be cracked
4. Like passwords
5. Encode
6. Many a download
7. A.T.M. need
11. English mathematician Alan, who was a key figure in the development of the modern computer and carried out important code-breaking work during World War II
12. Francis who developed a binary code in 1605 for use in cryptography
15. Number of digits in the decimal system
16. Rule for encoding or decoding messages

Answer on page 338

ACROSS

3. The best-known ones are at Giza

6. Inscribed stone found at the western mouth of the Nile in 1799 that permitted the deciphering of hieroglyphic writing

8. Person who copied important documents

9. Falcon-headed god regarded as the protector

Egypt, and thus one of the most important of the gods

11. Places to worship

12. Writing material prepared from the pithy stem of a water plant

16. Goddess of fertility, wife of Osiris and mother of Horus

17. Field that the Egyptians developed greatly, before

Pythagoras in Greece

18. Section

19. To supply water by means of channels

DOWN

1. They grow on the body and the head

2. Egyptian writing symbols

3. Egyptian ruler

4. French emperor who invaded Egypt in 1798

5. The Egyptians considered these honey-making insects very important

7. Crypt, vault

10. Husband of Isis and father of Horus

13. Creative activity such as painting

14. Site of the Great Sphinx

15. Important river for Egypt

Answer on page 338

ACROSS
1. Dangerous
5. Waste at a treatment plant
9. Poisonous
10. Inactive
11. Satchels
12. Rot
13. Back in the day
15. Trickle
16. Trash
17. Litter

DOWN
1. Homo sapiens
2. Relating to animals
3. Grim
4. Opposite of desired
6. Converting wastes into reusable materials
7. The human form
8. Kitchen sink device
11. A state of matter
14. Clean air org.
14. Scattered pieces of waste

Answer on page 338

ACROSS

1. Beyond the present
3. Put on
5. Central interest
8. Tendencies in societies and technologies
9. Surmise form these tendencies
12. Prognosticate
14. Remain
15. Toy with a tail
18. Anticipated moment in time when AI and will be so advanced whereby humanity will undergo a dramatic and irreversible change
20. They last for months

DOWN

2. Machinery and equipment developed from the application of scientific knowledge
4. Prophesize
5. Primary
6. Distress signal
7. Likely
10. Correct
11. Exam
13. In the cards
16. Tales of heroic exploits
17. Fourth dimension
19. To attempt

Answer on page 338

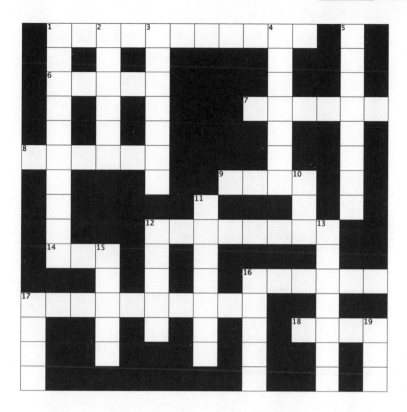

ACROSS

1. Scientific Study of Aging
6. Helicopter part
7. Recollection
8. Scattered
9. Stride
12. About time
14. Distress signal
16. Stress
17. People born and living at about the same time
18. Lateral

DOWN

1. Branch of medicine dealing with the health and care of older people
2. Stop working
3. Homes providing accommodations and health care for elderly people
4. Hereditary
5. Acquiring knowledge
10. Period
11. Inhaling cigarettes
12. Doctor
13. Foremost
15. Slumber
16. Slow moving mollusk
17. Chess or checkers, for example
19. Consume

ACROSS

1. Branch of medicine concerned with blood diseases
6. Phone purchase
8. Blood _____, for example A positive or O negative
10. Yellow-colored fluid in which blood cells are suspended
12. Instances
14. Color of blood cells that protect the body against both infectious disease and pathogens
16. Small metric unit of weight
17. Everything
18. Clumps of coagulated blood
19. Vessel that carries blood to the heart

DOWN

1. Opposite of cool
2. They carry blood from the heart to the other parts of the body
3. Red blood cells carry this from the lungs to the rest of the body
4. Bulk
5. Basic units of life, they can be white or red in blood
7. Cell fragments found in blood and involved in clotting
9. Smallest of the blood vessels; they exchange of substances between the blood and tissue cells
11. Red protein responsible for transporting oxygen in the blood
13. Blood _____, general term for veins, arteries, and capillaries
15. Chemical element Fe; essential for blood production

Answer on page 338

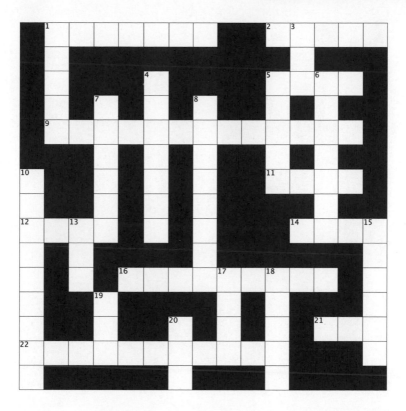

ACROSS

1. Sanitation
2. Opposite of dirty
5. Spot for a flame
9. Process of making something free from bacteria and other microorganisms
11. Stench
12. Secure
14. Scrub
16. Disease characterized by inflammation of the liver
21. Animal doc
22. It's next to godliness, in a saying

DOWN

1. Prehensile parts of the body
3. Floral wreath
4. Rest stop features
5. Washing need
6. Disease caused by infected water
7. Waste disposal system
8. A symptom of cholera
10. Sterilize
13. Opposite of near
15. State of being free from disease
17. Clock
18. Rubbish
19. Org. for doctors
20. Cover

Answer on page 339

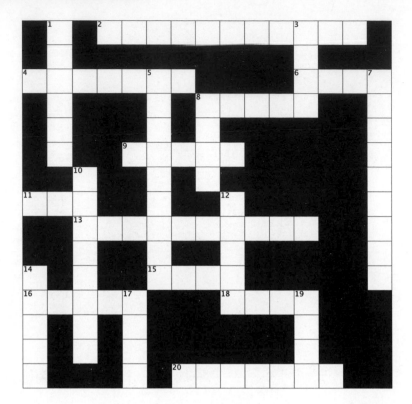

ACROSS

2. Inflammation of the tonsils
4. Windpipe
6. Doctor's tool for determining if the throat is infected
8. Climb onto
9. Sleep disorder
11. Color the throat might turn from a severe cold
13. Flap in the throat that prevents food from entering the windpipe and the lungs
15. Achy
16. Express
18. Discomfort
20. Blood _____, such as veins and arteries

DOWN

1. Voice box
2. Lab procedure
5. It connects the throat to the stomach
7. Inhaling and exhaling
8. Encounter
10. Enlarged lymphatic tissue between the back of the nose and the throat
12. Common throat ailment
14. Bird-related
17. Effortlessness
19. Void

Answer on page 339

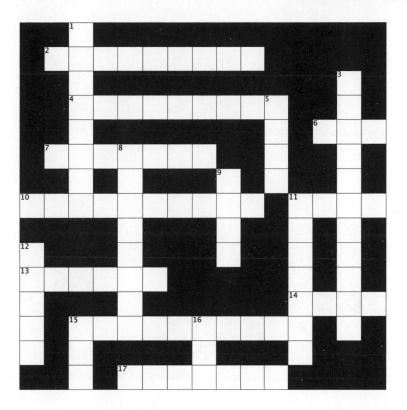

ACROSS

2. Museum studies
4. Objects with cultural or historical interest
6. Offer made at an auction
7. Arranged exhibition
10. Manhattan museum designed by Frank Lloyd Wright, with "the"
11. Decorative container
16. Famed Paris museum
14. Pottery and ceramics are made with this substance
15. Ancient city renowned for its library
20. The study of the past

DOWN

1. Organizing items for a museum exhibition
3. Washington, DC institution with 19 museums
5. Rescue
8. Keep in tact
9. Era
11. Center of Catholicism that's home to many museums
12. Opposite of newer
15. Museum holdings
16. Negative adverb

Answer on page 339

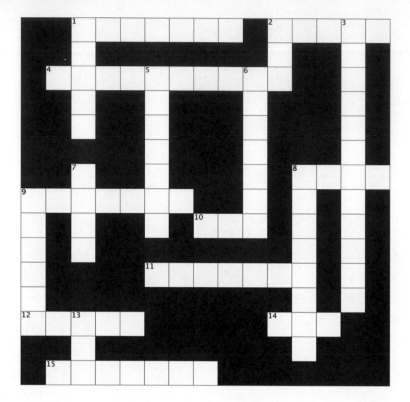

ACROSS

1. Meteorological conditions at any point in time
2. The final frontier
4. Gaseous envelope around Earth
8. Spray
9. Puffy clouds which look like cotton
10. Soft
11. It accompanies lightning
12. Opposite of big
14. Haze
15. In aviation, the measurement of the height of the base of the lowest clouds

DOWN

1. Life-sustaining liquid
2. Perceive
3. Moisture
5. Low altitude cloud
6. Arch of colors formed in the sky
7. Polluted haze
8. Type of cloud that forms from a nuclear explosion
9. Wispy, tufted cloud formation
16. Era

Answer on page 339

ACROSS

3. The sense of smell
6. Chemical signal between organisms
8. Smell is one of five
11. Adhesive paste
12. Balance used in labs
13. Like the smell of some flowers
15. Remember
16. Reptiles that smell with their tongues
18. Opposite of light
19. Scents

DOWN

1. Pain
2. Loss of the sense of smell
3. Smells
4. Sweet smelling stick that may be burned at an altar
5. Artificial fragrance
6. Like some smells left by animals
7. Aroma
9. The olfactory organ
10. Happiness or sadness, for example
13. Predatory marine fish with a keen sense of smell, especially for blood
14. Canines
15. Pleasant smell, fragrance, scent
17. Consumes

Answer on page 339

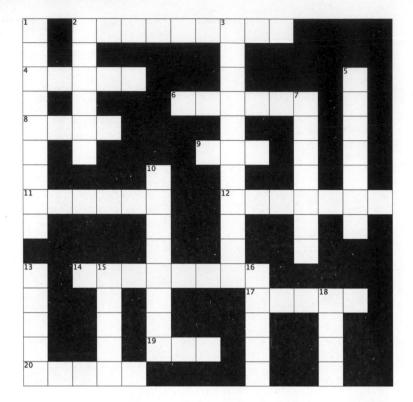

ACROSS

2. Schooling
4. Debate
6. Pedagogical ____, particular system of teaching
8. Wise
9. Enjoyable
11. Like some remote schools
12. Examining
14. Acquiring knowledge
17. Using knowledge in a practical way
19. Organ with a hammer
20. Disturbing sounds, which might hamper studying

DOWN

1. Place where schooling typically occurs
2. Involve
3. Teaching
5. Classes
7. Dedicated
10. Learn by heart
13. Teach
15. Assessment tools
16. Playful activities
18. To care very much

Answer on page 339

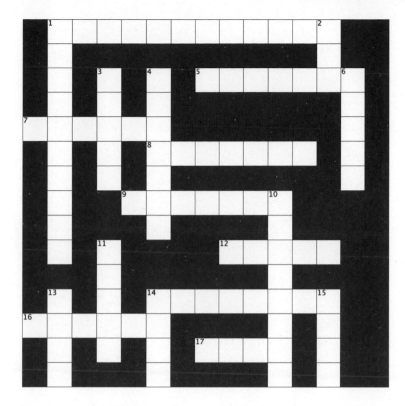

ACROSS

1. Condition in which the bones become brittle
5. Groups of cells in the body
7. They connect bones to each other
8. Mineral that builds bones
9. Fibrous bands that connect muscle to bones
12. Backbone
14. Bony
16. Soft, spongy tissue in the bone cavities where blood cells are produced
17. Back bone

DOWN

1. Branch of medicine dealing with bones, joints, and muscles
2. Perceive
3. What this puzzle is about
4. Tissues contracting together to produce a force
6. Head
10. Pain down the leg from the lower back
11. Squat
13. Opposite of soft
14. Whack
15. Extended

Answer on page 340

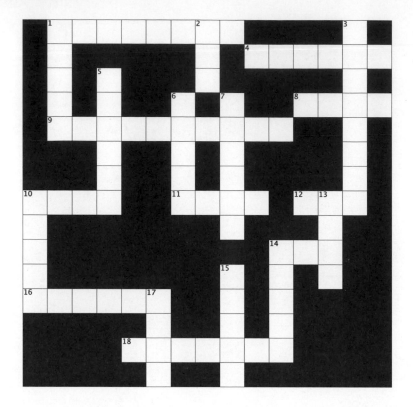

ACROSS

1. Medical treatment of feet and their ailments
4. Like a toenail infection
8. Like feet before a wedding
9. Ability to move from one place to another
10. Foot's digits
11. Footprint maker
12. IV units
14. _____ Majesty
18. Painful swellings on the big toes

DOWN

1. Foot levers used for powering a bicycle
2. A primary color
3. Hardened parts of the skin
5. Curved parts of feet
6. There are 26 of these in each foot
7. Part of the calf
10. It's below the femur
13. Foot ailment
14. Back part of the human foot below the ankle
15. Claw of a bird of prey
17. Boot accessory

Answer on page 340

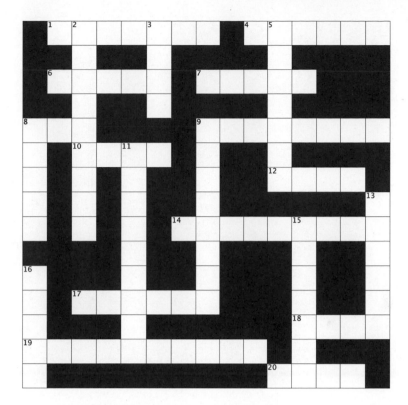

ACROSS

1. Relating to earthquakes
4. Slight earthquake
6. Alternate rising of falling of the sea
7. Violent shaking caused by a tremor or full-blown earthquake
8. Perceive
9. Like Earth's plates
10. Celestial explosion
12. Pour
14. The Richter scale measures this, for an earthquake
17. Deluges
18. Large, heavy, thick-furred mammal
19. Highest point

DOWN

2. Point where an earthquake originates
3. Large amount
5. Numerical scale for expressing the magnitude of an earthquake
8. Quake
9. Enormous sea waves caused by earthquakes
11. Erupting mountains
13. Trepidation
15. Small earthquake
16. Planar fracture, the most common cause of earthquakes

Answer on page 340

ACROSS

3. Large marine animals found in in similar areas of the world as seals

6. Seals belong to this class of animals

8. Opposite of "land"

9. Sound emitted by sea lions, similar to that of dogs

10. Forelimbs that seals use for steering and propelling themselves in water

12. Large marine animals distinguished by two large downward-pointing tusks

17. Arthropods such as crabs and lobsters

18. Strategy

19. Upkeep

DOWN

1. Opposite of large

2. Fat of seals

3. Bristles growing from the face that seals use as sensors

4. Marine

5. Predatory marine mammals who circle their prey

7. Region

10. Makeup of schools

11. Cultivate

13. Region around the North Pole

14. Epic

15. Killer whale

16. It might be left after a wound heals

Answer on page 340

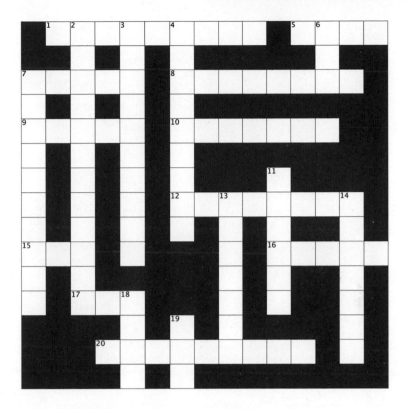

ACROSS

1. Practices or theories concerned with the supernatural
5. Warning sign
7. Greek philosopher and student of Socrates, who wrote the *Republic*
8. Float in the air, magically
9. Playing cards used for fortune-telling
10. In Greek mythology, the first mortal woman sent to Earth by Zeus with a box of evils
12. Trance-like state in which people are highly prone to suggestibility
15. Move at a fast speed
16. Opposite of outer
17. Sixth sense, briefly
20. Astrological chart

DOWN

2. Ability to see the future
3. Unconventional
4. Thought transference
6. Encountered
7. Ancient Greek mathematician who promoted numerology
11. Group of signs, include Aries, Virgo, and Libra
13. Anyone considered to have powers of clairvoyance
14. Doubter
18. Nudge
19. _____ of gold

Answer on page 340

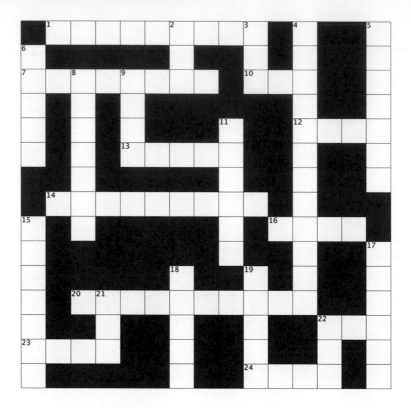

ACROSS

1. Respiratory illness
7. Ability to resist a specific infection
10. Place to soak
12. Stand
13. Infectious viral disease that causes internal bleeding, occurring mainly in Africa
14. Inflammation of the liver
16. The common _____
20. Malaria, West Nile virus, and dengue fever are spread by these insects
22. Understand
23. Final
24. Fall

DOWN

2. Consume
3. Performance
4. Infectious bacterial disease caused by the growth of nodules in lung tissues
5. Severe vomiting and diarrhea caused by drinking contaminated water
6. A pathogen
8. Rubeola
9. Epithet
11. Contagious viral disease of animals that causes convulsions and can be transferred to humans by bites
11. Sexually transmitted disease
15. German measles
17. Pest
18. Contagious viral disease that causes swelling of the salivary glands
19. _____-19
22. Drink slowly

Answer on page 340

ACROSS

2. Snake or scorpion danger
6. Science of poisons
8. Infectious disease caused by highly toxic bacteria; a toxic powder can be made from it
9. A salt of hydrocyanic acid that is extremely toxic that was used infamously in the Jonestown tragedy
12. Standard
13. Ailment
14. Medicine that counteracts a particular poison
16. Toxic metal found in seafood
17. Type of oil, and a plant whose seeds can be toxic

DOWN

1. Highly toxic protein extracted from the seeds of the castor-oil plant
3. Some of these usually edible fungi are highly toxic
4. Highly poisonous compound obtained from the *nux vomica* plant that is thought to have poisoned Alexander the Great
5. Unusual
7. Plant with tonic and medicinal properties, native to East Asia
10. Chemical element As, which is used in medicine, but can also be used as a strong poison
11. Poisonous evergreen shrub characterized by clusters of white, pink, or red flowers
15. Giant

Answer on page 341

ACROSS

1. New York, Tokyo, and London, for example
4. Opposite of rural
6. Income
8. People
9. Urban ____, process of developing and designing urban areas
13. Like green energy
15. Upkeep
16. Equivalent
17. Sequence

DOWN

1. Neighborhoods
2. Urban problem
3. Like the area outside of a city
5. Regions
7. Urge forward
9. Declarations of assurance
10. New
11. Positive development
12. Interested groups or people
14. Severe

Answer on page 341

ACROSS

1. Timber
6. Outermost portion of a stem or branch
7. Tree's stalk
8. Tall deciduous tree with jagged leaves
10. Plant
11. Conifer from which fragrant, durable timber is made
13. Arboreal plants
15. Advantage
16. Part of a trunk or branch that has been cut off
18. Caring
19. Evergreen coniferous tree, with needle-shaped leaves and has cones

DOWN

1. Opposite of found
2. They jut out from trees
3. A primary color
4. Trees that are usually found in broad-leaved temperate and tropical forests
5. The science of trees
9. Reddish-brown timber from a tropical tree, used typically to make furniture
10. Common Christmas tree
12. Dark timber from a dark hardwood tree
13. Lumber
14. Tree that bears acorns
17. Hold

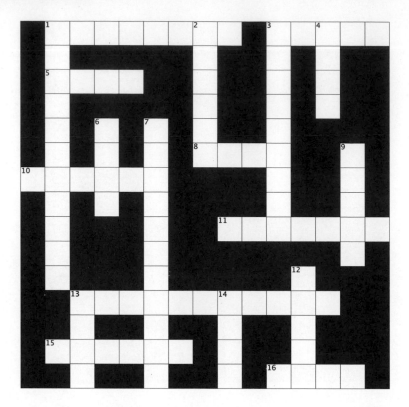

ACROSS

1. Science of soil management and crops
3. Animals that are destructive to crops
5. It provides the staple diet for half the population of the world
8. Opposite of west
10. Flora
11. Inherited
13. Tilling of the land
15. Southern crop
16. More than ninety percent of rice is produced on this continent

DOWN

1. Science and practice of farming
2. Animal dung, used as fertilizer
3. Contamination
4. Earth
6. Terrestrial
7. Refurbishment
9. Apples, oranges, or lemons, for example
12. Provisions
13. Plant cultivated as food
14. Region

Answer on page 341

ACROSS

1. Branch of medicine concerned with skin
6. Skin layer
7. Blisters, lesions
9. Indoor animal
10. Skin flakes
12. Appearance
14. Lump in the throat
15. Dermatitis
16. Fibrous protein forming the outer skin layer
19. Make contact with the hand or fingers
20. Autoimmune skin disease marked by red, itchy patches

DOWN

2. Skin surface
3. They bite
4. Develop
5. These are caused by eczema or psoriasis
8. Color
9. Human being
11. Dermal, referring to the skin
13. Sore on the skin
17. Blow
18. Pimples

Answer on page 341

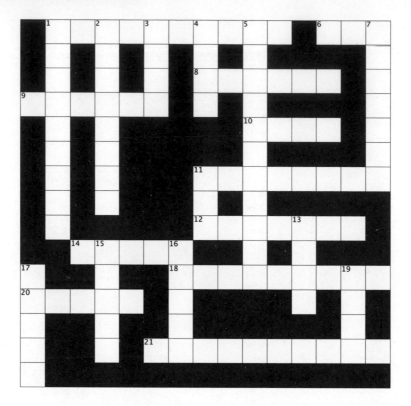

ACROSS

1. Branch of medicine concerned with liver, gallbladder, and pancreas
6. Opposite of on
8. Organ that acts as a blood filter
9. Tumor-producing disease
10. Yellowish-brown fluid produced by the liver to aid digestion
11. Machine that replaces malfunctioning kidneys
12. Stomach
14. Workload
18. A serious liver disease
20. Lower parts of the ears
21. Operation in which a patient is given a new healthy organ

DOWN

1. Inflammation of the liver
2. Organ of the digestive system
3. A drop of fluid from an eye
4. Opposite of gain
5. Small sac-shaped organ beneath the liver, which stores bile
7. Cleans
11. Deoxyribonucleic acid
13. Earth's natural satellite
15. Stadium
16. Scrub
17. Hemoglobin
19. Fe, chemically

Answer on page 341

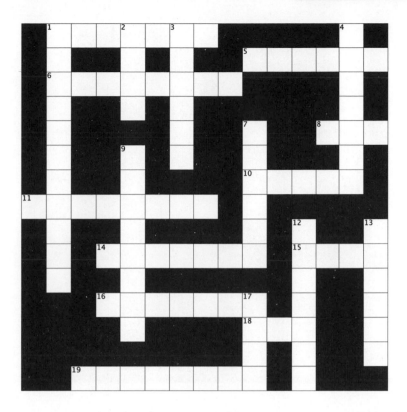

ACROSS
1. The study of muscles
5. Flat
6. Might
8. Feline with strong leg muscles
10. Impetuous
11. They are made of amino acids
14. They connect muscles to bones
15. Stiff
16. Referring to the heart
18. Backtalk
19. Decrease in size

DOWN
1. Muscular state
2. Scallion
3. Look with wide-open eyes
4. Abdomen
7. Muscle ____ control forces moving through the body
9. Like the body's framework
12. Degeneration of muscles
13. Opposite of weak
17. Coagulate

Answer on page 342

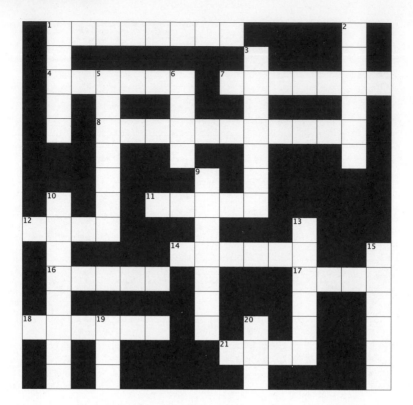

ACROSS

1. The science of growing fruit
4. Fruits that are said to "keep the doctor away"
7. Curved, long yellow fruit, which is peeled before it's eaten
8. Farming
11. They are planted
14. Fruit named for its color
16. Oval, typically purple, fleshy fruits
17. Small green fruit often used as a garnish
18. Type of fruit high in vitamin C
21. Soft pear-shaped fruits with sweet flesh and many seeds

DOWN

1. Yellowish-green fruits, narrow at the stalk and wider at the base
2. Fruits used to make wine
3. Containers that cider may be sold in
5. Round stone fruits with pinkish-yellow skin
6. Fruit layer
9. Small, often round, juicy fruits without a stone
10. Popular Halloween gourd
13. Large, pulpy, round fruits
15. Common citrus fruits
19. Color of cherries
20. Stone of a fruit

Answer on page 342

ACROSS
1. Plants, collectively
4. Grain used for food
5. Territory
7. It is eaten by horses
10. Front yard
11. Unwanted wild plant
11. Garden with grass
14. Matches
15. Grown mainly to make bread
17. Slender-leaved grass that grows in water
19. Grass grown as a grain; wheat-like
20. Grain that can be made into a meal
22. Meadow

DOWN
1. Panorama
2. Color of grass
3. Opposite of wide
6. Places for growing vegetables
8. Leaf of grass
9. Dried stalk of grain
11. It is needed by the grass to survive
12. Woody grass that grows in the tropics
13. They fall in the fall
16. Region
18. Stain

Answer on page 342

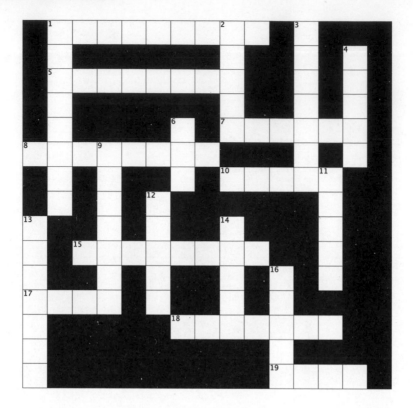

ACROSS

1. Branch of science and medicine concerned with hearing
5. Units indicating the intensity of a sound
7. Reverberations
8. Spiral cavity in the inner ear that produce nerve impulses in response to sounds
10. Opposite of outer
15. Buzzing or ringing in the ears
17. Stopper
18. It vibrates in response to sound waves
19. Fleshy part at the end of the ear

DOWN

1. Referring to the sense of hearing
2. Some are noble
3. Equilibrium
4. Disturbing or meaningless sound
6. It is secreted in the ear canal for protection
9. What the ears permit
11. Respond
12. Pathway from the outer ear to the middle ear
13. Eardrums
14. Opposite of inner
16. Hearing-related

Answer on page 342

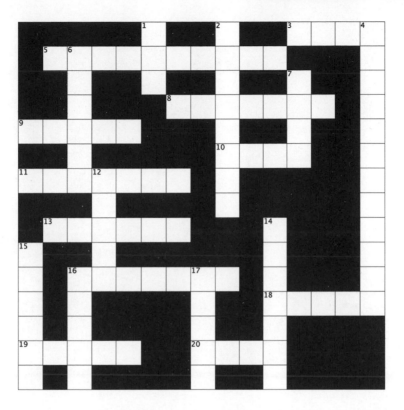

ACROSS

3. Nightly celestial body that bore many mythical meanings in antiquity
5. The science of the past
8. Pictographic or alphabetic, for example
9. Basic type of social grouping
10. Metal that lends its name to a prehistoric age
11. Relics, remnants
13. Metal that lends its name to a prehistoric age
16. Written accounts
18. In the Stone Age they were made from stone
19. Remnants, residues
20. Vital early human discovery

DOWN

1. Lair
2. Early humans
4. Extinct human species
6. Objects surviving from an earlier time
7. Unknown; abbr.
12. Age that preceded the Bronze Age
14. Garments
15. People who do not have a permanent abode, moving around regularly
16. Guidelines
17. Outline

Answer on page 342

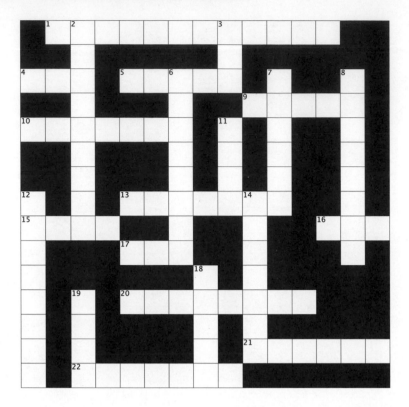

ACROSS

1. Genetic _____, is the process of altering the genetic makeup of an organism
4. Delay
5. Frothing at the mouth
9. An organism or cell that has been produced to be genetically identical to another one
10. Hereditary
13. Pathogens responsible for many communicable diseases
15. Valentine's flower that has been the object of genetic engineers
16. Deoxyribonucleic acid
17. Ribonucleic acid
20. Gene processing
21. Complete set of genetic information in an organism
22. Relating to moral principles

DOWN

2. Living things
3. Help
5. The first organisms to be modified in the laboratory
7. Organisms such as flowers and shrubs
8. Doctor's study
11. Hearing things
12. Mating
14. Gene _____, is a method allowing scientists to change the DNA of many organisms
18. Thin protuberances that project from the cell body, especially in single-celled organisms
19. Common rodents

Answer on page 342

ACROSS
1. Branch of medicine dealing with the treatment of cancer
7. Development of secondary malignant growths at another place from the primary site of cancer
8. Cancerous
10. Collection
11. Like neoplasms, which may be benign or malignant
13. Association
16. Revoke
17. Growths that may be benign or malignant
20. Cancer that starts in the bones or soft tissue

DOWN
1. *Carmen* or *Pagliacci*, for example
2. Cancer-causing
3. The first number
4. Type of fungal infection
5. Facts
6. Food regime
7. Gender most effected by prostate cancer
9. Med. group
12. Skin cancer
14. Blood cancer
15. Surgical extraction of cells or tissues for examination
16. Emit
18. Cancer treatment, briefly

Answer on page 343

ACROSS

1. The science of insects
5. Consume
6. Nocturnal flying mammals
7. What a mosquito does
9. Of the water
10. What mosquitos inject into their prey before and during blood-sucking
12. Pond-dwelling amphibians
who feed on mosquitoes
14. Metal in brass
16. Musical acuity
17. The biting mosquito is this gender
18. Liquid mosquito food
19. Mosquitoes have six of these

DOWN

2. Webs designed to keep mosquitoes away
3. Mosquito-borne disease that causes vomiting, tiredness, fever, tiredness, and headaches
4. Immature form of mosquitoes
5. NYC hrs.
6. Feathered, winged vertebrates
8. Inflammation of the brain, caused by an infection
or an allergic reaction
11. Scratchy
12. Opposite of near
13. Identical
14. Virus transmitted by mosquitoes causing fever and rash, identified originally in Africa
15. _____ fever is a virus transmitted by mosquitoes, affecting the liver and kidneys
17. Household pests

Answer on page 343

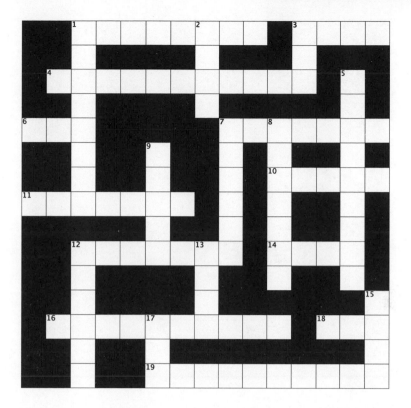

ACROSS

1. Guns, rifles, and pistols, collectively
3. Round bullet
4. Missiles fired from a weapon
6. Handheld firearm
7. Projectiles that can be propelled to a great height and distance
10. Spikes
11. Automatic gun
12. Ammo
14. Opposite of far
16. Path followed by a projectile
18. Corp. bigwig
19. Bullet supply

DOWN

1. Relating to criminal investigation
2. By habit
3. Motor coach
5. Military weaponry
7. Firearms fired from the shoulder
8. Mortars
9. Severe
12. Straight shooting tube of a gun
13. Discretion
15. Midday
17. Epoch

Answer on page 343

ACROSS

4. Dimension
6. Large amount
9. Ample
10. It is measured in hours, seconds, days, months, etc.
13. Capacity
14. Total amount
15. Try
16. It can be measured in Fahrenheit or Celsius
19. Large quantity
20. Era
21. Total

DOWN

1. Measuring system based on the meter, liter, and gram
2. Fixed share
3. Heaviness
5. Approximately
7. Arrange neatly
8. Quantity
11. Nothing
12. Count
15. Entire
17. Heap
18. Perpetually

Answer on page 343

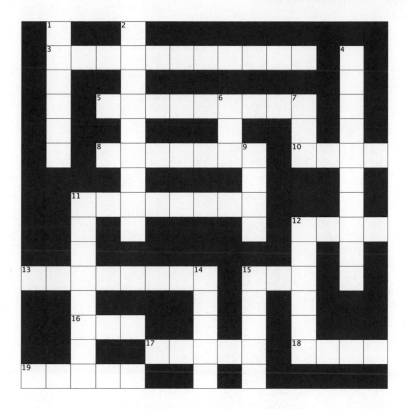

ACROSS

3. Back and forth movement at a regular speed
5. Rate at which a vibration occurs
8. They vibrate when plucked
10. Rhythmic backward and forward movement
11. Weight suspended from a pivot, so that it can swing freely
12. Heat
13. Resistance encountered by an object when moving over a surface
15. Net
16. Racket
17. Sign of life
18. Rebounding sound
19. Spoken

DOWN

1. Vibrations that travel through air
2. Pulsating
4. Ancient mathematician who studied the vibrations of strings; known mainly for his right-triangle theorem
6. Unit of energy
7. Opposite of no
9. Equivalent
11. Type of wave with a repeating continuous pattern that defines its wavelength and frequency
12. Move unsteadily
14. When scratched against a chalkboard, they produce a skin-crawling effect
15. They transfer energy from one place to another

Answer on page 343

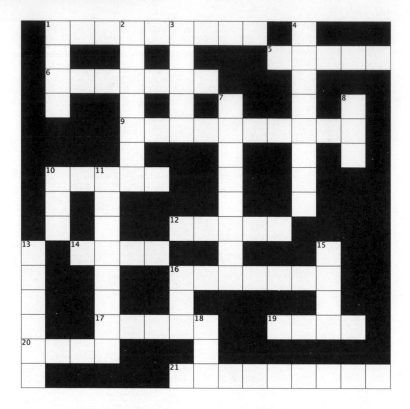

ACROSS

1. The study of wave frequencies
5. Musical sounds
6. Actuality
9. Musical sounds that seem to be out of tune
10. It can be high or low
12. Rudimentary
14. American genre of music based on syncopation and improvisation
16. Musical era of Bach and Vivaldi
17. Rate of speed of a musical piece
19. Choir part
20. Sacred
21. Musical era of Haydn, Mozart, and Beethoven

DOWN

1. Musical instrument, with a graduated series of parallel strings, played by plucking
2. Pleasant-sounding
3. Symbols denoting musical sounds
4. Musical era of Chopin, Brahms, and Tchaikovsky
7. Person who writes music
8. In music, it can be in the major or the minor
10. Music genre for Madonna or Lady Gaga
11. Music quality
13. Musical time pattern
15. What you might clap on
16. Be-____, popular style of jazz developed in the 1940s
18. Loosen

Answer on page 343

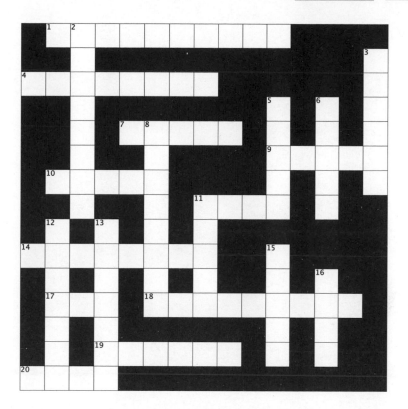

ACROSS

1. What is seen in the mirror
4. The science of shapes and figures
7. What a mirror reflects
9. What appears in the mirror
10. Exam at the end of a course
11. Disturbance that carries energy
14. Writer
13. Wafer in which ice cream is put
17. Automobile
18. Greek god who fell in love with his own reflection in a pool
19. U-shaped
20. Prank

DOWN

2. Conceited
3. They can be acute, obtuse, right, etc.
5. Lewis Carroll's female protagonist who falls through a mirror
6. Mirror material
8. An optical one tricks the eye into seeing something other than what it is
11. Liquid that has the property of reflection
12. Bowl-shaped
13. Make light change direction when it enters at an angle
15. Rush
16. It may be bad if you break a mirror

Answer on page 344

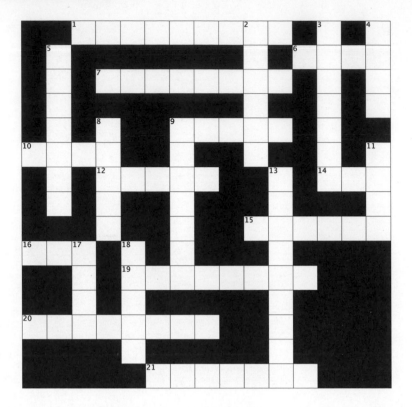

ACROSS

1. Size, extent
6. Chess, checkers, or backgammon, for example
7. Branch of mathematics studying figures and shapes
9. Timepiece
10. What an older clock does
12. Self-evident truth in geometry
14. Age
15. Takeoff
16. Animal doc
19. Living thing
20. Unit made up of two or more atoms
21. Malady

DOWN

2. Straight ahead
3. Possible cause of a virus
4. Opposite of right
5. Branch of science in which vectors are studied
8. Hockey player's need
9. Generating a genetically identical copy of a cell or organism
11. Arithmetic
13. Conveys
17. Trouble's partner
18. The "F" in F = ma

Answer on page 344

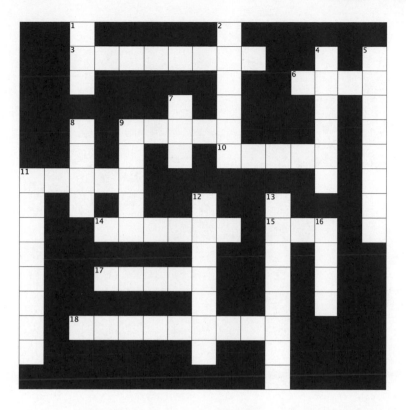

ACROSS

4. Snowflakes, for example
6. Polar inhabitant
9. Color of fresh snow
8. Measure of snow, as in "two ____ of snow"
10. Form
11. Enhance
14. Solid
15. Opposite of high
17. Slushy snow
18. Snow mass of snow falling rapidly down a mountainside, growing in size as it falls
20. "____ snow," slushy snow
21. Ice pellets mixed with snow

DOWN

1. Frozen water
2. Snow fall
4. Small ice particle
5. Wandering
7. Popular Christmas tree
8. This puzzle's theme
9. H_2O
11. Severe snowstorm
12. Thawing
13. Bodies of snow that compress into enormous, thickened masses
16. It blows

Answer on page 344

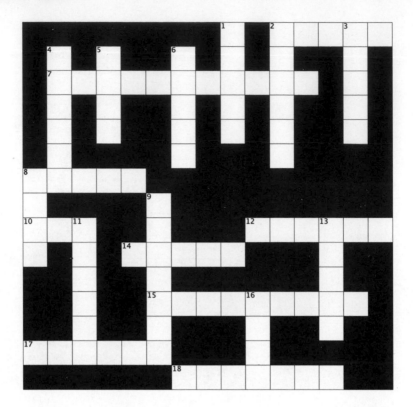

ACROSS

2. Ice crystals formed on the ground when the temperature falls below freezing
7. Process of thickening
8. Get through to
10. The Baltic, for one
12. Covering at the poles
14. Pieces of ice added to a drink to make it colder
15. Six-sided
17. Crowded
18. Floating mass detached from a glacier

DOWN

1. Melted ice
2. Iced
3. Winter precipitation
4. Spike found hanging in winter
5. It falls in winter
6. _____ ice, dangerous
transparent coating on a road
8. Break
9. Like the ice in some beverages
11. North pole
13. Dairy product
16. Liquidly lump

Answer on page 344

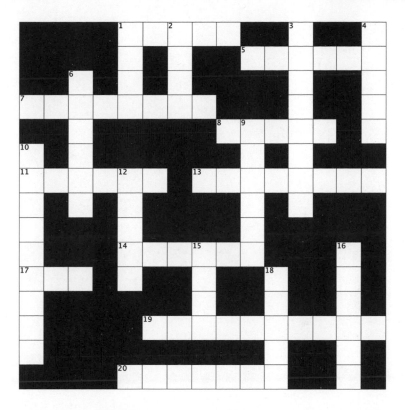

ACROSS

1. It is equal to equal to 16 ounces
5. Body mass
7. Amount
8. Opposite of small
11. Liveliness
13. Ability to float in water
13. System of measure
17. Fuel
19. Einstein's theory
20. Things working together cohesively

DOWN

1. Basic unit of the avoirdupois system
2. Standard measure
3. About 2.2 lbs
4. The building blocks of the universe
6. Physical substance
10. Tiny unit of weight
12. Lab weights
15. Fact-based
16. The "M" in MA
18. Twosomes

Answer on page 344

ACROSS

1. A characterization of density;
5. The "m" in $E = mc^2$
7. Force that keeps things from flying off the Earth
8. A body's relative mass
11. Standard
12. Greek scientist and mathematician who shouted "Eureka," running naked through Syracuse, after discovering a principle concerning buoyancy, at least according to legend
16. Capacity
17. Apiece
18. Plumage
19. Iron or steel, for example

DOWN

2. Amount
3. Fundamental unit of all matter
4. A particular exemplar of matter
6. Lungful
9. Score received on an exam
10. Opposite of thinness
13. The tops of sea waves
14. What Archimedes purportedly shouted after discovering a law of buoyancy
15. Stay above water
18. What birds can do

Answer on page 344

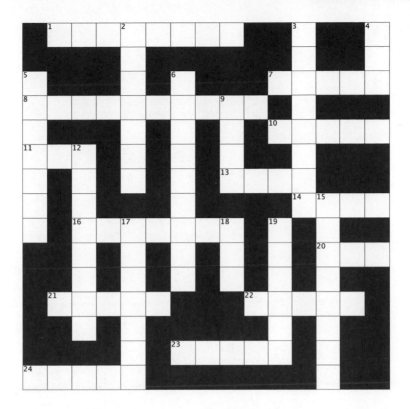

ACROSS

1. Release
7. Lasers amplify this
8. Method of generating three-dimensional images
10. Building blocks of matter
11. Account
13. Den
14. Out of harm's way
16. Slicing
20. Glimmer
21. Laser lights
22. Construct
23. Neither gas nor liquid
25. It is used by pilots to determine direction, speed, location, etc.

DOWN

2. Procedure that may use a laser
3. Laser or inkjet computer accessory
4. Univ. outside of Boston
5. Basic units of light
6. Emission of energy
9. Get better
12. Machine-readable codes in the form of numbers and parallel lines
15. Jet
16. Device for grooming or gardening
17. Fuel for cars
18. Neither a solid nor a gas

Answer on page 345

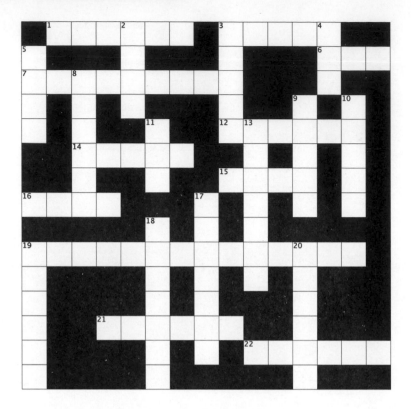

ACROSS

1. The Milky Way, for example
3. _____ System
6. Conclusion
7. Large northern constellation between Perseus and Pegasus; named after an Ethiopian princess of Greek mythology
12. The Milky Way's shape
14. Type of star that has escaped the gravitational pull of its home galaxy
15. Ring of light
16. _____ Major, also known as the Big Dipper
19. Group of stars forming patterns traditionally associated with mythological figures
21. Greek messenger of the gods
22. Group of stars said to represent a swan that was the form adopted by Zeus

DOWN

1. Building block of all matter in the universe
3. Luminous astronomical bodies
4. Color of Mars
5. Big _____ Theory
8. Relatively small stars
9. Conspicuous constellation, said to represent a hunter holding a club
10. _____ hole, a region of space with an intense gravitational field
11. Star which the Earth orbits
13. Mars, Mercury, and Venus, for example
17. Italian Renaissance astronomer who discovered the four of the moons orbiting Jupiter
18. Type of very dense, compact stars thought to be made up primarily of these subatomic particles
19. Middle
20. Rather than red, Mars is closer to this color

Answer on page 345

ACROSS

1. Metals made by combining two or more metallic elements
3. Shiny metal with the symbol Ag
5. Pliable
6. Red-brown metal, used for making electric wires
8. The most precious of metals
10. Chemical which, when combined with chlorine, produces salt
11. Utters
15. Funerary ash container
17. What this puzzle is about
16. Opposite of cold
19. Yellowish-brown alloy of copper that gave the name to the age after the Stone Age

DOWN

1. The most abundant metal in the Earth's crust with the symbol Al
2. Sturdy and strong alloy of iron
4. _____ Man
7. Silvery-white metal with the symbol Pt
8. Precious stones
9. Copper or aluminum, for example
12. Burning
13. Opposite of northern
14. Containing iron
16. Kind of acid

Answer on page 345

ACROSS

4. One of the three states of matter
5. Middle
7. They are constructed by termites
8. Equal
10. Road caution
12. Spherical body
13. Our planet
16. Apollo 11's destination
17. Substantial
19. Spheres used in sports
20. Possesses
21. Half the globe

DOWN

1. North and South
2. Borders
3. Branch of mathematics that studies space
6. Half diameter
9. A gemstone
11. Spherical celestial bodies orbiting the Sun
14. Spheres of soap
15. It's 360°
18. Opposite of thin

Answer on page 345

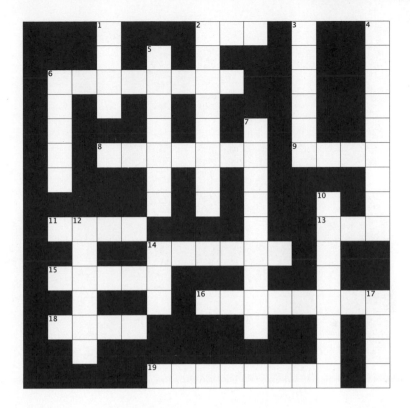

ACROSS

1. Carbon dioxide, for example
6. Like a substance found in a lab
8. The "O_2" in CO_2
9. A carbon _____, is a celestial body that contains more carbon than oxygen
11. Fuel that is mostly carbon
14. Carbon _____, determines the age of organic matter from the relative proportions of the carbon isotopes that it contains
15. The Carbon _____ describes the process of carbon atoms traveling from the atmosphere to the Earth and then back
16. Porous black carbon solid, used for drawing and barbecuing
18. Brown, soil-like boggy material, made partly of decomposed vegetable matter
19. They are charted on the periodic table

DOWN

1. Coal, oil, or natural gas, for example
2. Crystalline form of carbon, used in pencil tips and in lubricants
3. Metal mixtures
4. Carbon _____, measures the impact our activities have on the environment
5. Precious stone of pure carbon; used in jewelry, especially rings
6. Top layer
7. Plant that returns every year
10. Variants of chemical elements with the same number of protons and electrons, but a different number of neutrons
12. It combines with carbon to form carbon dioxide
14. Lair
17. Existence

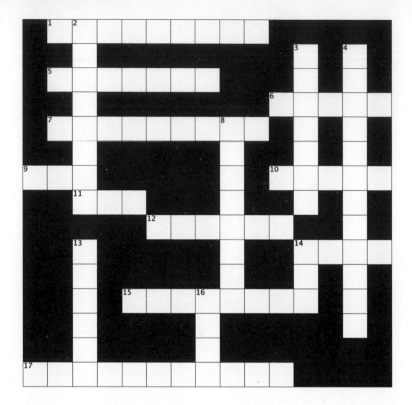

ACROSS

1. Monochrome
5. Radiator fluid
6. Life-sustaining fluid
7. Combustible
9. Hydrogen, for example
10. Luminous celestial bodies consisting mainly of hydrogen and helium
11. Earth orbits it
12. Subatomic particle with a positive charge
14. They emanate from the Sun
15. Subatomic particle with a negative charge
17. Methane is an example of this compound

DOWN

2. Unscented
3. They orbit the Sun
5. Hair removal from the body using an electric current
9. Hydrogen, compared to the other elements
14. _____ hydrogen, a common rocket fuel
16. Scorch

Answer on page 345

ACROSS
1. Monochrome
8. Lungful
9. Envelope of gases surrounding the Earth
12. Oxygen _____, an enclosure that aids a patient's breathing
13. Fume
15. Endeavor
16. Oxygen-16, Oxygen-17, and Oxygen-18, for example
19. Oxygen storage vessel
20. Like some gases
22. Breathing

DOWN
2. Unscented
3. Living things
4. Cleansing substance
5. Color of the sky
6. Oxygen _____, placed over the mouth and nose to aid breathing
7. Responsive
10. Oxygen compounds
11. Luminous celestial bodies
14. Thin shell on the outside of the Earth
16. Very unpleasant
17. Layer of the stratosphere that absorbs the Sun's ultraviolet radiation
21. Source of solar energy

Answer on page 346

ACROSS

1. Relatively inert
6. Scream
7. Oxygen compound
10. Nitrogen _____ is the processes whereby nitrogen passes from the atmosphere to the soil and to organisms and back into the atmosphere
11. Consume
12. Nitric _____, corrosive and poisonous liquid with strong oxidizing properties
14. Fragrance-free
17. Most common noble gas
19. One of the three states of matter
20. Gustatory sense

DOWN

2. Item in the periodic table
3. Freon, for one
4. Ogle
5. Nitrogen is vital to these living things, which grow from the ground
8. Versions of the same element, but with different numbers of neutrons
9. Poison that contains a carbon-nitrogen bond
12. Pungent gas, a compound of nitrogen and hydrogen, used as a cleaning fluid
12. Gustatory sense
13. One of the three states of matter
15. Beach memento
16. Static
18. One of the three states of matter
19. Distress call

Answer on page 346

ACROSS

1. Looks
5. Sodium chloride
6. Just made
8. Gustatory sense, which some additives aim to enhance
9. Protect
11. Oil's partner
12. Pigmented additive
14. Food additives used to stabilize processed foods
16. Conclusion
17. Noxiousness

DOWN

2. Jams
3. Tastes
4. Substances used to counteract the deterioration of food products
7. Coffee or tea drinker's request
9. Stevia or equal, for example
10. Like a substance found in a lab
13. Opposite of bases
14. Consume
15. Plant

Answer on page 346

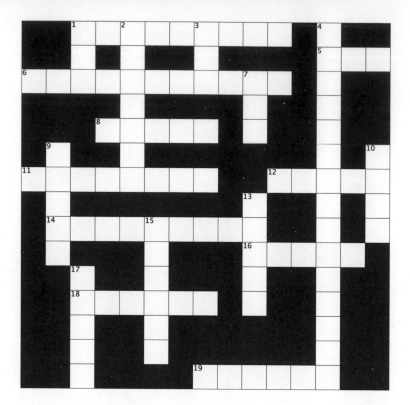

ACROSS

1. Franklin's main field of study
5. Ribonucleic acid
6. Acknowledgment (which came later for Franklin)
8. Franklin is also well-known for her study of the structure of this common pathogen
11. Form of carbon used commonly to make pencil tips
12. Franklin's work on defeating this illness
14. Famous British university
16. Opposite of small
18. Scientist James, a discoverer of the double helix structure of DNA, for which he received a Nobel Prize, argued that Franklin should have received it as well
19. Sagely

DOWN

1. Public health org.
2. Franklin's nationality
3. Collection
4. Science of crystals, in which Franklin also excelled
7. Be in debt
9. Scientist Francis, who shared the Nobel Prize for the double helix model of the DNA
10. Combustible black rock found mainly in underground deposits and used as fuel—Franklin's work on this substance was renowned
13. Double _____, the structure of DNA
15. Justification
17. Medal

Answer on page 346

ACROSS

5. Latin word for gold, from which its chemical symbol is derived
6. Gold's color
8. Gold or silver, for example
10. Pliable
11. Gold seeker
14. Fine particles
17. Prized
18. Pet with a short memory
20. Relating to finance

DOWN

1. Flexible
2. Necklaces and rings, for example
3. Money
4. Aristocratic
7. Stately tree
9. A gold one of these is given for first place
12. Gold _____, frenzy that brought many people west
13. Pre-chemistry science that attempted to convert base metals into gold
14. Title
15. Natural-occurring pieces of gold
18. Headpieces, often of gold, worn by kings and queens

Answer on page 346

ACROSS

3. Valuable
5. Silver or gold, for example
6. Objects of adornment, often made of silver
7. Color between white and black
9. A silver one can kill a werewolf, or symbolize a solution to a difficult problem
12. Thwart
16. Silver is the traditional gift for the celebration of a twenty-fifth _____

DOWN

1. Utensils
2. Piece of tableware
3. Trophy
4. Currency
5. Ductile
8. Silver _____, solid formerly used in photography
10. A silver one symbolizes a hopeful aspect to a difficult situation
11. Glittery
13. Actual
14. Roman poet known for the *Metamorphoses*
15. desire strongly
19. Fact-based

Answer on page 346

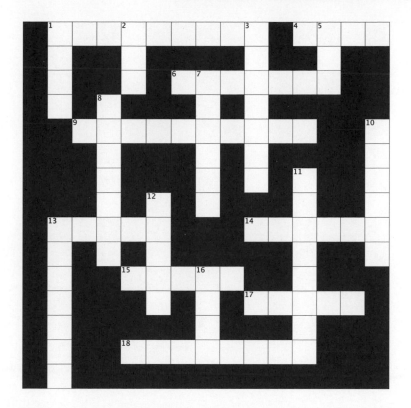

ACROSS

1. _____ nervous system, works automatically, without a person's conscious effort
4. Disposition
6. Messages
9. _____ nervous system, outside the brain and spinal cord
13. Transmitter of electrical impulses
14. Type of neurons that reflect the behaviors of others
15. Referring to the connective tissue of the nervous system
16. Backbone
18. Minute gaps between nerve cells, where electrical signals are converted into information

DOWN

1. Threadlike part of a nerve cell where impulses are transmitted to other cells
2. The first number
3. _____ nervous system, the brain and spinal cord
5. Photo _____
7. Breathe in
8. Web-like system
10. Nerve tissue strands
11. Neural _____, neurons interconnected by synapses to form larger brain networks
12. Biological units
13. Nerve cells
16. Zone

ACROSS

4. Animal without a backbone
7. Ovum
9. Opposite of short
11. Earthworms respire through this
13. Harvests
14. Pharmaceutical
18. Like an earthworm's shape
19. Fertilizing mixture

DOWN

1. Arms and legs
2. Crawling
3. Earthworm's home
5. Tunneling
6. Organs that earthworms do not have
8. Positive quality
10. Small lizard
12. Night crawler
15. Operate, like a computer program
16. Gunk
17. Lure

Answer on page 347

ACROSS

1. Water movements
5. Opposite of near
8. Slow, regular movement of rolling waves
10. Celestial body that effects the tide
13. Rivulet
15. Equipment
16. Mythical singer who lured sailors to their deaths
17. Location
18. English physicist Sir Isaac, who explained that tides were the product of the gravitational attraction of celestial bodies
19. Sea _____

DOWN

2. Opposite of falling
3. Flow out
4. Adriatic and Mediterranean, for example
5. Opposite of rise
6. Dangerous type of tide
7. Deluges
9. Opposite of strong
11. Tidal wave
12. Force that attracts a body toward Earth's center
14. Revolution
15. Italian scientist that gave an explanation of tides in 1632
17. Stage

Answer on page 347

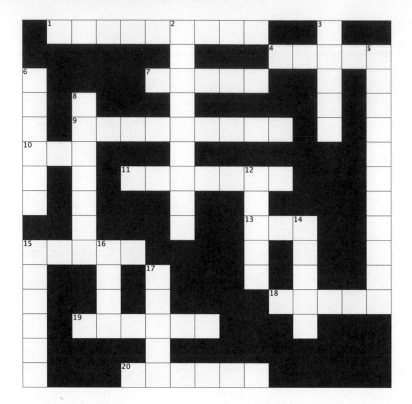

ACROSS

1. The _____ Galaxy is visible to the unaided eye, sharing various features with the Milky Way system

4. Small constellation, named after the ram whose Golden Fleece was sought by Jason and the Argonauts

7. Heavenly bodies

9. Space science

10. Negative particle used with verbs

11. Ancient astronomer, among the first to study the constellations in the *Almagest*

13. Large constellation, named to represent the lion slain by Hercules

15. Planet in the solar system named after the Roman goddess of love

18. Runs

19. Constellation representing a bull that was tamed by Jason

20. Astrological chart

DOWN

2. Collection of stories belonging to one tradition

3. Constellation representing balance

5. Constellation said representing a centaur with a bow and arrow

6. Northern constellation, representing a flying swan

8. Arrangement

12. _____ Way, galaxy that contains the solar system

14. Constellation said to represent a hunter with a club and shield

15. Whirlpool

16. _____ Minor, constellation that contains the polar star, Polaris' it means "little bear"

17. Constellation said to r representing goddess of the harvest

Answer on page 347

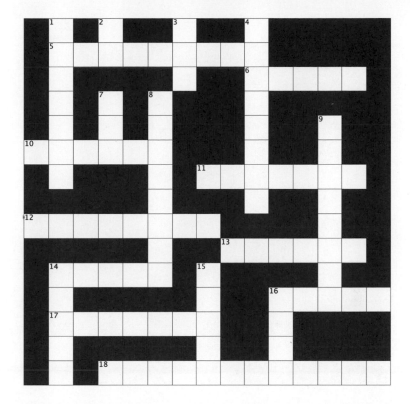

ACROSS

5. Greek philosopher who invented the syllogism
6. Obvious
10. English scientist Sir Isaac, who saw science and mathematics as interconnected
11. Mathematical relationship expressed in symbols
12. Amount
13. Is, mathematically
14. Ruler of a hive
16. Benefit
17. Arithmetic with letters
18. The mathematics of triangles

DOWN

1. Renaissance Italian scientist who characterized mathematics as the language of nature
2. The ratio of the circumference of a circle to its diameter
3. Opposite of high
4. The mathematics of points, lines, triangles, squares, etc.
7. Collection
8. Expression showing the relationship among variables
9. Subject with derivatives and integrals
14. Subatomic particle with a fractional electric charge
15. Greek philosopher who founded the Academy, where he taught arithmetic, geometry, music, and astronomy
16. Fundamental unit of matter

Answer on page 347

ACROSS

1. Study of ancient writing
6. Biblical gift
7. Remnants
8. Prefix referring to the stars
9. Wide-eyed in wonder
11. Type of pyramid with flat platforms
14. Documents, texts written by hand
17. Slab of clay or stone used for making inscriptions
18. Slumber

DOWN

1. Manuscript page, from which the content has been taken off so that the page can be reused for another document
2. Study of the origin of words
3. Study of past cultures, with methods such as excavation of sites
4. Writings
5. Carbon _____, used to determine the age of organic matter
10. Writing symbols
11. Commandment verb
12. Platter
13. Squanders
15. Everything
16. Ceremonial ritual

Answer on page 347

CARL JUNG (1875-1961)

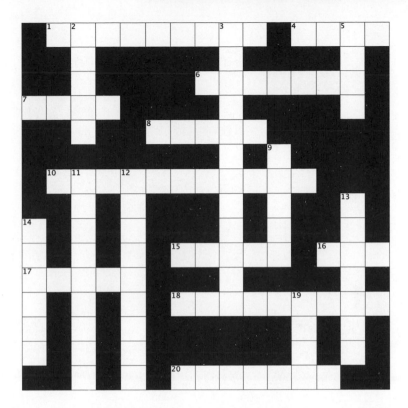

ACROSS

1. Jung's notion of a primitive mental image inherited from our ancestors

4. Jung saw these ceremonial face coverings as symbolic

6. Pessimistic

7. Get better

8. Ancient stories, such as those of the gods, which Jung saw as psychological descriptions

10. Someone's distinctive character

15. Jung's nationality

16. A creative activity, such as painting

17. Scent

18. Instinctive sense

20. They represent important cultural things

DOWN

2. Unwind (which Jung believed to be essential)

3. Jung's profession

5. A person's essential being

9. Heavenly bodies

11. Opposite of introvert

12. Feeling or perception

13. The character we present to others

14. Jung's archetype for negative aspects of character that remain unconscious

19. Labor

Answer on page 348

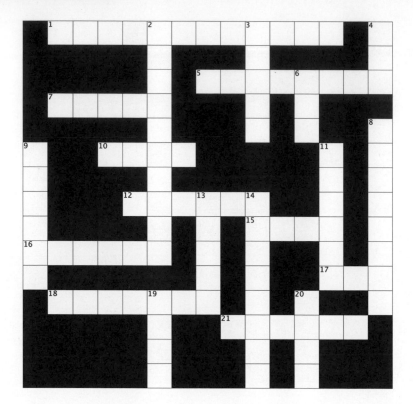

ACROSS
1. Handling
5. Tallying
7. Tactile sense
10. Jewelry for a finger
12. Fingers
15. Verb preceder
16. Small
17. Star orbited by Earth
18. Finger bone
21. System of arithmetic based on 0 and 1

DOWN
2. Capable of grasping
3. Shortest finger
4. Excavate
6. The end of a finger
8. Finger _____ drawing with the fingers
9. Third finger
11. Unique part of the finger that is used by forensic investigators to identify a culprit
13. Pointer
14. Action with the thumb and the middle finger
19. During
20. Scratching need

Answer on page 348

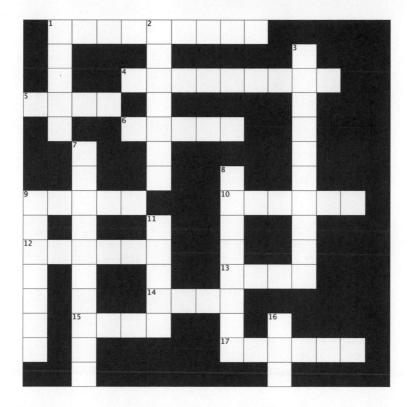

ACROSS

1. Greek philosopher who invented the logical syllogism
4. General inference or conclusion
5. Disposition
6. Method of showing that something is true, as in geometry
9. Entail
10. Think things out logically
12. Understand by instinct
13. Signify
14. Plunge
15. Remain
17. Einstein's _____ of relativity
18. Method of collecting information

DOWN

1. Self-evident proposition
2. Pythagorean _____
3. What is inferred at the end of a logical argument
7. Supposition requiring further research
8. Coherent statement of an idea
9. Emulate
11. Investigate something
16. Collection

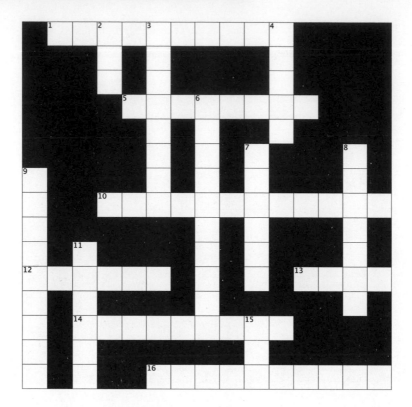

ACROSS

1. Initial theory or supposition
5. Simulating
10. Notes or opinions
12. Do again
13. Confident
14. Proficiency in an area
16. Use data in a misleading fashion

2. Face value
3. What scientists put forth
4. Marks
6. Test
7. Unverified assertions
8. Corroborate
9. Fact-based
11. Oral communication
15. Drai

Answer on page 348

ACROSS
1. Estimation
4. Calculate
5. Raw information
6. Mark registering a number
8. Type of bar graph
11. Based on figures
13. Estimate
15. Assess
16. Rational

DOWN
2. Likelihood
3. Boost
7. Drawback
9. Trial
10. Take in
12. Add up
15. Opposite of new

Answer on page 348

ACROSS

1. Large, high-speed computer: _____ frame
5. Intel product
7. Portable computer
8. Web address
10. RAM part
11. Excursion
12. Instruments
14. Unwanted messages on the Internet
15. A group of computers using common communication protocols
18. File with small bits of data that is used to identify a computer
19. Like most music recordings today
21. Streaming annoyance
22. Computer bug
24. Cyber criminal
25. Standalone malware program
26. Problem

DOWN

1. Handheld computer pointing device
2. Sick
3. Sodium
4. Small personal computer
6. Code writer
9. Digital storage units
12. Number of digits in a binary system
13. Computer input devices
16. Appliance requiring toner or ink
17. With "up," slow speed internet
20. Retreat
23. Feminine pronoun

Answer on page 348

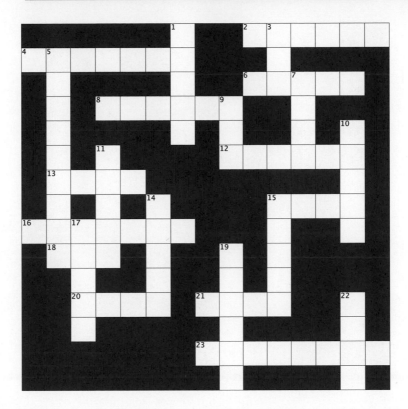

ACROSS

2. Mallet
4. Appliance or device
6. Instrument filled with liquid used to determine if a surface is perfectly flat
8. Twisty hardware store purchase
12. Tool used for turning nuts and bolts
13. Wheel turner
15. Electric cord
16. What a knife is used for
18. Call for help
20. Traffic marker
21. Metal tool with jaws, used to hold hard objects firmly in place
23. A spherical metal ball suspended on a string so that it can swing back and forth

DOWN

1. A rigid bar on a pivot, used to move or lift heavy objects
3. Log splitter
5. Technical equipment
7. Futile
9. Tool with teeth
10. Circular object that allows a car to move
11. Some Vegas games
14. Instrument with a blade
15. Wood splitter's need
16. Clasping device holding things together
17. Portable means of illumination
19. Sculptor's tool
22. Adhesive substance

Answer on page 349

ACROSS

1. Like a device that needs to be plugged in
7. Node with a large number of links
8. Online Public Accessing Catalog (*abbr.*)
9. Word before media, for Facebook or Twitter
11. Connecting networks
15. Closer
16. Balance
17. Digital program over the Internet, similar to the radio
19. YouTube upload
21. Spotify or iTunes offering
22. Format for compressing digital images

DOWN

1. Kindle purchase
2. Authorize
3. Devices
4. Opposite of new
5. Co. leader
7. Auxiliary digital memory
10. Online place to visit
13. Handheld products from Apple or Samsung
14. Data packets used to identify and access a server or user.
17. Pictures, for short
18. Facts

Answer on page 349

ACROSS

1. Field that integrates computer science and engineering
6. Car, for short
7. Celestial body that is seen on a clear night
8. Pea holder
9. Together
10. Rank Order List (*abbr.*)
11. The branch of science and technology involved in machine design and structure
13. Unpleasant sound
14. Comp. memory
15. Clay figure brought to life in Jewish legend
16. Mass. research university, known for its work in robotics and other new technologies
17. Like many robots or movie aliens
19. Opposite of front
20. Leonardo da ____, who designed early robots

DOWN

2. Defective
3. Give instruction to
4. Electronic devices used to navigate the Internet, among other things
5. Part person, part robot
6. iPhone alternative
8. Scheduled or automated
10. Laptops, televisions, and other devices
12. A spirit imprisoned in a bottle
14. ____ control
18. PC alternative

Answer on page 349

ACROSS

1. Branch of science dealing with the manipulation of atomic particles
6. Opposite of analog
8. Computer storage unit
10. Online correspondence
12. Nav. aid
13. Human-resembling robot
16. "Deep _____" was a chess-playing computer; named after a common color
17. Common pronoun
19. Kind of developer
21. Photo often taken with a smartphone
23. Most used search engine

DOWN

2. Connected to the Internet
3. Computer component
4. Disturbing sound
5. Essential product
7. Amazon's virtual assistant
8. Milliliter (abbr.)
9. Large computer storage unit
11. String of symbols or characters

for accessing a computer system
14. CD part
15. Sending a message via smartphone
18. YouTube offering
19. "_____ AI;" also refers to a plant's unit of reproduction
20. Online journals
21. Start of something
22. Romantic feeling

Answer on page 349

ACROSS

4. Type of force that effects charged particles
7. Coffeeshop convenience
9. Gesture of approval
10. Opposite of hot
12. Episodic dramas, which started out on radio
14. Balanced
15. X

16. "_____ radio," program in which people express their opinions
19. Main offering of iTunes or Spotify
21. "Wait Wait ... Don't Tell Me!" airer
22. "_____ radio," type of radio broadcasting baseball, football, and tennis, among others

DOWN

1. Transmitting radio programs
2. Information program
3. Early radio
5. Guglielmo, who was considered "the father of the radio"
6. Objective
8. Data processed by computer

11. Semiconductor device capable of amplification
12. Radio programs that rely on an ongoing story
13. Amplitude Modulation
17. Frequency Modulation
18. Armed conflicts
20. Co. leader

Answer on page 349

ACROSS

1. Picture
5. Trend, style
7. The first number
10. Photography of conflicts between nations or states
12. Appearance
14. 1950s White House nickname
15. Candid
17. Very thin
18. Photography using computer technology
19. Misdeeds, offenses, recorded by CSI photographers
20. Payment due for a service, such as photography
22. Device for taking photos
23. The art of melody and harmony

DOWN

1. Handheld device
2. Opposite of on
3. "Camera ____," literally "dark room," a box in which an image is projected inside. It is the predecessor of the modern camera
4. Software company with a Creative Suite used for creating images
6. Company that was the leader in film-based photography, founded by George Eastman
8. Leonardo da ____ was among the first to work with the camera obscura, a predecessor of the camera
9. Photo often taken on a phone
11. Photography often featured in National Geographic
13. Photo and video sharing social media service owned by Facebook
16. Photograph of a person
19. Opposite of less

Answer on page 349

ACROSS

1. Applied sciences
6. Storage of information, making it ready for delivery
8. Catchy melody
10. Film
11. Audio medium, pioneered by Guglielmo Marconi
12. Publish
13. Information
15. Home entertainment device
19. Opposite of in
20. Some daily publications
22. Vimeo upload

DOWN

1. Not *that*
2. Systems for sending messages
3. Head gesture conveying agreement or greeting
4. Global computer network
5. Music alternative
7. Web address
9. Periodical publications containing articles and illustrations
13. Information
14. Popular streaming service
16. Tech that enables phone calls over the internet
17. _____ opera, program in serial form
18. Irritating sound
21. Ampersand meaning

Answer on page 350

ACROSS

5. E-mails, phone calls, and text messages, for example
8. Word with media or land
9. Word in some Commandments
10. Convert a document into digital form
12. Composing text
14. Combine or join together
15. Org. that put a man on the moon
17. "Snail _____," postal system
18. A tall, rounded vase
20. Primal emotion of terror or fright
21. Oration
23. Addressee
24. Opposite of always

DOWN

1. Rotate
2. Photographs
3. Speak
4. Snog
5. Television station
6. The contents of communications
7. Correspondent
11. Electrical impulse sent or received
13. Body sign or message, mainly with the hands
16. Pictures used in digital messages
18. Sell or where a sale happens
20. Prepare for something unpleasant
23. Writing instrument that is "mightier than the sword"

Answer on page 350

ACROSS

3. Distributing audio and visual content
6. Imaginative story or narrative
7. Watch online
9. *American* _____, long-running competition show
10. A test of knowledge, as on television _____ shows
12. CBS and NBC, for example
14. Car or kitchen need
15. Physical means used for communications
19. Artificial body orbiting the Earth that can be used for broadcasting
20. Big initials in computers
21. Place where an artist works

DOWN

1. Game of Thrones airer
2. Television by subscription
4. Companies that produce movies and television
5. Popular streaming service
7. Shortly
8. Relax
10. Target audience for Daniel Tiger or Sesame Street
11. Earth orbiter that can be seen on a clear night
13. _____ TV, "unscripted" television programs
15. Type of necklace
16. Basic units of matter, consisting of a nucleus and orbit particles
17. *Seinfeld* or *The Big Bang Theory*, for example
18. Opposite of day

Answer on page 350

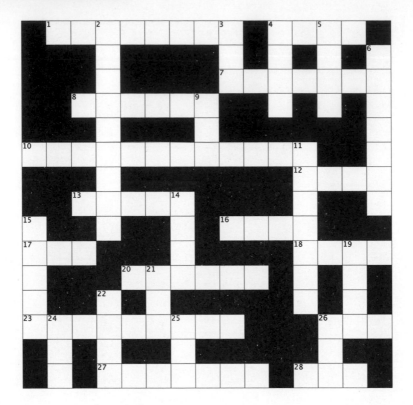

ACROSS

1. Coded software instructions to control the operations of computers
4. Price
7. Software designed to damage computers
8. Program that controls the operation of a computer device.
10. Codes that define a computer program
12. Twelve months
13. Entered
16. "On the _____"
17. Computing acronym
18. Opposite of difficult
19. Yield
23. Information stores
26. Nineteenth-century mathematician and computer pioneer Lovelace
27. Computer programs allowing users to enter or alter text
28. US spy agency

DOWN

2. The "O" in "OS"
3. Total amount
4. Icy
5. Unwanted messages sent on the Internet
6. Computers that manage access to networks such as the Internet
9. "Stuck in a _____"
11. Procedures
14. Try
15. iPhone alternative
19. Spoken
21. Chant heard at the Olympics
22. Word after video or ball
24. Smartphone purchase
25. Important HS exam
26. Interface that defines interactions between multiple software applications, briefly

Answer on page 350

ACROSS

1. Python, Java, and HTML, for example
6. Smartphone purchase
7. A program that converts instructions so that they can be executed by a computer
8. Strike, like a key on a keyboard
12. Implement or instrument
13. Programs used by a computer
15. A step, code, or sequence in a computer program
20. Precious metal, whose symbol is "Au"
21. Ada, who was a pioneer in the development of modern computers
22. The final step in a computer program
26. Unit of computer memory

DOWN

2. Computer programming instructions for carrying out calculations or other operations
3. Undo, unfasten
4. "Video _____," played mainly on the Internet
5. Facts
9. The "P" in "HTTP"
10. A program that is repeated over and
11. Xbox enthusiast
12. English mathematician Alan, who is famous for his code-breaking work during World War II
14. Easy-to-use computer programming language based on familiar English words, designed for beginners—hence its name
16. Courage
17. Set of computer program instructions
19. Opposite of in
20. Break into a computer or a computer network
22. Assemble

Answer on page 350

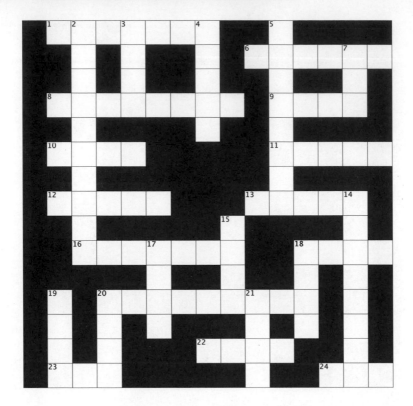

ACROSS

1. Opposite of artificial
6. Machines that resemble humans and capable of carrying out some human tasks
8. Collection of data and instructions for organizing it
9. Design
10. Care
11. Contemplate
12. Game that ends with "checkmate"
13. Scarecrow's want
16. Signs, such as computer icons, that stand for something
18. Close
20. An instruction that is loosely defined, designed mainly to assist in doing something
22. "____ learning algorithms;" sophisticated algorithms contrasted with "shallow learning" ones
25. Search Engine Optimization (abbr.)
26. ____ Today

DOWN

2. Sets of computer-usable instructions, organized in a step-by-step fashion
3. Quantity chosen as a minimal standard
4. Acquire knowledge or skill
5. A mechanical brain, figuratively
7. Number of digits in the decimal system
14. The first nuclear-powered submarine, 1954 (US Navy)
15. Utilizes
17. "Deep ____," was the name of a chess-playing computer
18. Ill
19. Some internet trolls
20. Protagonist
21. A diagram with "branches" that is used commonly in the design of programs

Answer on page 350

ACROSS

1. Connected to the web
5. Letters in a URL
7. Opening
8. Set of rules governing the transmission of data between devices
10. Opposite of in
11. Assistance, help
12. Everything, the whole quantity
14. Searches for, online
19. News deliberately meant to deceive

DOWN

2. System of connected computers
3. Most used online encyclopedia
4. Free online dictionary
5. Worldwide
6. Internet addresses under the control of a particular organization or person
9. Exchanging or portioning files with other users
13. Forwarding packets from one network interface to another
15. Basic quantity; the number one
16. Shape, structure, manifestation
20. Letters in a URL
21. Anxiety about being excluded, for short

Answer on page 351

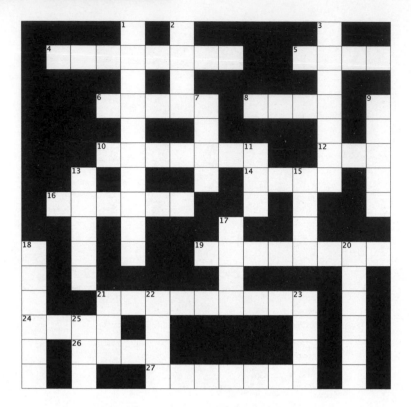

ACROSS

4. Zuckerberg's social networking website

5. Anxiety about being excluded, for short

6. Twitter posting

8. Upcoming, following

10. Personal information on a social media platform

12. Info found in an "about me" tab of a website

14. Pleasant, agreeable

16. Cellphone, in some countries

19. Computer framework for sharing information

21. Social networking service centered on photographs

24. Hammer or screwdriver, for example

26. Appeal

26. Confidentiality

DOWN

1. Connecting with users on social media

2. Timbre, sound

3. Online video-sharing platform

7. One of two who share a womb

9. A picture word, such as the smiley

13. A discussion board on the Internet

15. Cook

17. Online journal

18. Short video-sharing social networking service

20. Website where you earn karma

21. Sick

22. _____chat, multimedia messaging app

23. A large number

25. Decide

Answer on page 351

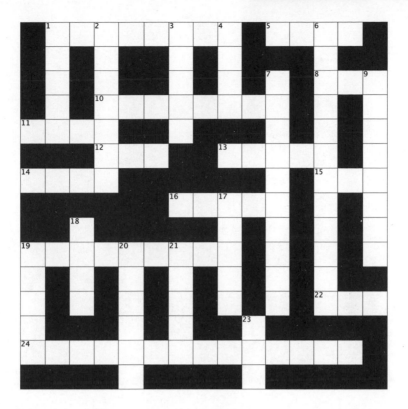

ACROSS

1. Written or recorded communications
5. Raw information
8. Corporate V.I.P.
10. Broadcast, send out
11. Identical
12. Univ. near Harvard
13. Prefix relating to the stars
14. Information about current events
15. Texter's reaction to something funny
16. Truths
19. Converted into a code
22. Perceive
24. Propaganda meant to mislead

DOWN

1. Press
2. Groups of related programs
3. Bold, ambitious
4. Unsolicited or illegal email messages
6. Skills
8. The science of analyzing data mathematically
9. "Information ____," excessive information
17. Instructions for computer programs
18. A fraudulent scheme
19. Finished, completed
20. Longs for
21. Off-limits
23. Diagram of an area of land or sea

Answer on page 351

ACROSS

1. William _____, writer who spread the term "cyberspace" in his novel *Neuromancer*
4. Community
5. Popular Thanksgiving pie
6. As they say, *this* is mightier than the sword
9. Machine many people use daily
11. Place to live
12. Relating to the entire world
14. Uncover a password or code
16. Lighter-than-air craft, often used as a toy
17. Like virtual reality
19. Playing card with a single spot or image
20. Gaining unauthorized access to a computer system
22. Equine animals

DOWN

2. Of sound mind
3. Name of the novel, by William Gibson, that made the term "cyberspace"
4. Bigwig
5. Sending fraudulent emails to dupe users to reveal personal information
7. Large Australian bird
8. System of connected computers
10. Dress
13. Assaults
15. Like a video that spreads rapidly on the Internet
18. Send forth; make a sound
21. Age

Answer on page 351

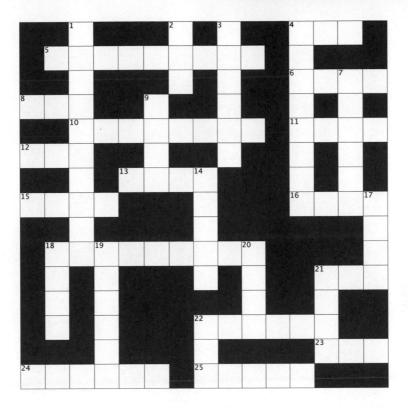

ACROSS
4. Part of a URL
5. Popular online encyclopedia
6. Online journal
8. Many a Twitter troll
10. System of interconnected computers
11. Capri or Skye, for example
12. Corp. bigwig
13. Turn over
15. Snare, pitfall
16. Start-up money
18. The "H" in "HTTP"
21. Owed
22. Packet of data used to subsequently access the same server
23. Cannabis, colloquially
24. Computer that manages a network
26. Auxiliary memory system for high-speed retrieval

DOWN
1. Popular online dictionary
2. Tim Berners-_____, an inventor of the internet
3. Connected on the internet
4. Internet locations
7. Connected to the internet
9. Say
14. Summons, before cellphones
17. Go out with
18. Computer language
19. Ivan who developed the theory of conditioning by working with dogs
20. Popular order at a Mexican restaurant
21. _____ Learning, as opposed to Shallow Learning
23. Org. concerned with public health

Answer on page 351

ACROSS

1. Computations
5. Stages
6. Arrangement of words to create meaningful structures
7. Identical
9. Something to solve
12. Opposite of high
13. Opposite of in
14. Pleasant
15. Line of reasoning
17. Sudden attack
18. Head, as a committee
21. Kind of engine used for locating sites on the web
23. Arranging data in a prescribed sequence on a computer
24. Notoriety

DOWN

2. Manufacturing with machines
3. Steps, directions, as in a computer program's code
4. Prefix expressing negation
8. Diagram representing a process
10. Number system consisting of 0's and 1's
11. Ancient geometer, who anticipated the notion of algorithm in his *Elements*
16. English mathematician and computer scientist Alan, who was instrumental in developing the modern notion of computer algorithm
18. Opposite of open
19. Handle
21. Opposite of on

Answer on page 351

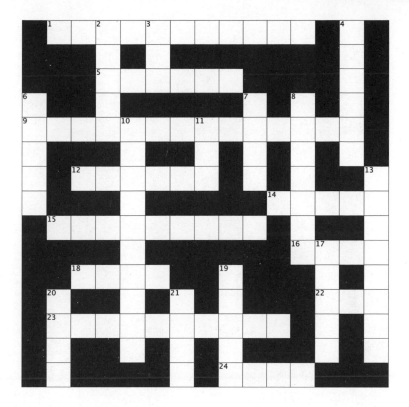

ACROSS

1. Facts
5. Opposite of hero
9. Categorization
12. Diagrams
14. Illegal offense
15. Like "x" in an equation
16. Opposite of dangerous
18. Appraise
22. The "P" in RPM
23. Amount in each hundred; proportion
24. Every one

DOWN

2. Actual data
3. Uncooked
4. Quantity
6. Number of points in a game
7. "_____ news;" misleading, deceptive information
8. Numbers
10. The mathematical science of data
11. Org. known for audits
13. Systematic investigation and collection of facts
17. Put something to practical use
19. Phase
20. Opposite of closed
21. Standard quantity

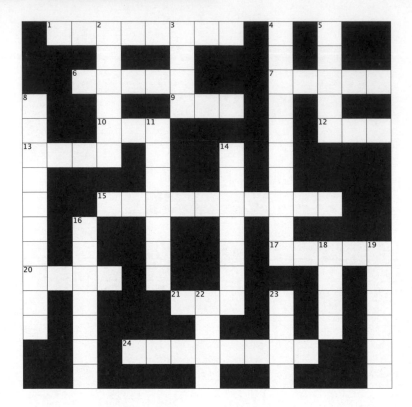

ACROSS

1. System of connected computers

6. Images or ideas that get repeatedly shared online

7. Like a video that spreads broadly on the Internet

9. Opposite of no

10. Like some projects, briefly

12. Early computing acronym

13. Trends

15. Keeping in contact, interacting on social media

17. Opposite of always

20. _____ study, research on a particular person or group

22. Unwell,

24. Audio-based downloads

DOWN

2. Tendencies

3. Curious

4. Routing

5. Connected

8. Those who've found Instagram fame

11. Video-sharing website

14. "_____ Reality;" computer simulation of an environment

16. #

18. Conceited

19. Site where users share content and earn karma

22. Delicate fabric

23. Music, literature, and painting, for example

Answer on page 352

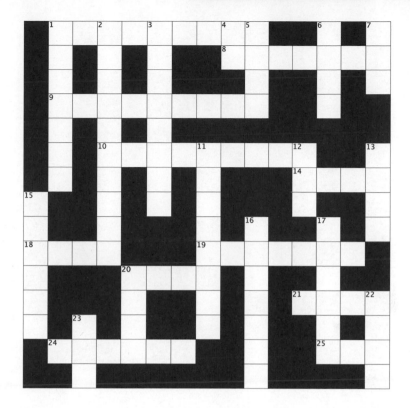

ACROSS

1. Dependency
8. Basic literacy skill
9. No longer close to someone
10. Capacities
14. Aid
18. Spur, stimulate someone into action
19. Apprehension, unease
20. On the _____; that is, immediately
21. Noticed
24. Well-being
25. Plant seeds

DOWN

1. Memory loss
2. Incapable of concentrating
3. Maximum amount
4. Either's go-with
5. Requirement
6. Care
7. Era
11. Secluded
12. Timid
13. Period of time
15. Rely on
16. Investigate, search for something
17. State of emotional strain
20. The immaterial part of a being
22. The latest information
23. One of the "W's" in WWW

Answer on page 352

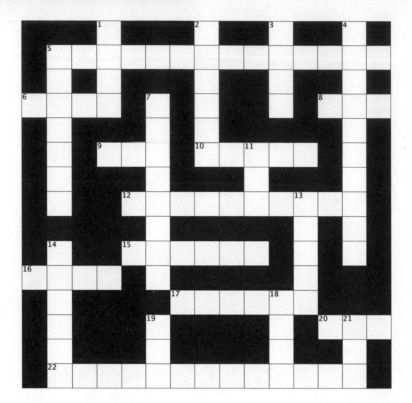

ACROSS

5. Propaganda meant to deliberately deceive
6. Highest ranking playing cards
8. Inquire
9. Autonomous computer program that can interact with other users and programs
10. Narrow piece of paper
12. Biased or false information (especially political)
15. Opinion or idea based on trust, or formed over time
16. Anxiety about exclusion, for short
16. Opposite of "yes"
17. Hearsay
20. Not new
22. Theories that cannot be proven in a controlled study

DOWN

1. Utilizations
2. Scams
3. "_____ news;" misleading, deceitful information
4. Theory or belief that an event was caused by a covert group
5. Swindle, cheat
7. Online world
11. Manage or arrange something fraudulently
13. Labels
14. Spread a rumor
18. Drive
19. Opposite of good
21. An untruth

Answer on page 352

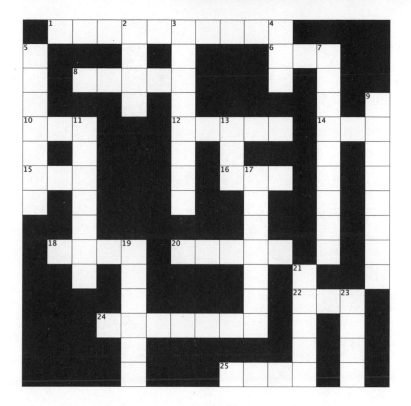

ACROSS

1. Some Sony or Samsung products
6. Common street name
8. Medium of newspapers, books, magazines, etc.
10. Opposite of new
12. Electronic mass communications medium that became dominant in the 1920s
14. Before the present
15. Uncover (as in investigative reporting)
16. Night before
18. "____ media;" media that reach large audiences
20. Prefix meaning "many"
22. &
24. Exact
25. Prejudice

DOWN

2. Long, heroic poem
3. Cyberspace
4. Opposite of old
7. Periodical publications, usually with illustrations and devoted to particular subjects, such as sports or fashion
9. Get from the Internet
11. Opposite of analog
13. Owed
17. Word before reality
19. ____ media, such as Facebook or Twitter
21. Truths
23. Unorganized information

Answer on page 352

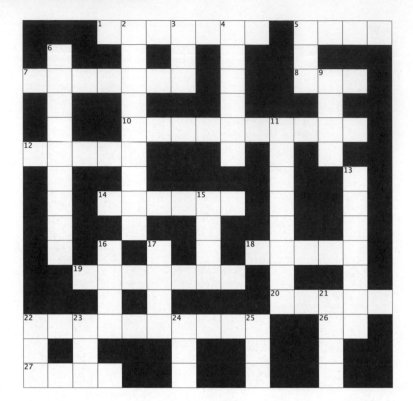

ACROSS

1. The "V" in "VR"
5. Identical
7. The "R" in "VR"
8. Ovum
10. Sequenced computer instructions, designed to seek information
12. Android
14. Representations that are intended to show how something should work
18. Opposite of darkness
19. The Looney Tunes, for example
20. Environmental
22. Those trained to go into outer space
26. Symbol for argon
27. A cassette or reel

DOWN

2. Replica, emulation
3. Plaything
4. Attraction
5. Perceive
6. Replicate
9. Video _____
11. Instructing
13. Meteorological conditions
15. Building block brand
16. Matching set of two
17. Stupefy
20. A negative adverb, as in "Do _____ do this"
21. Every one
22. Painting or drawing, for example
24. Peak
25. Smartphone purchase
26. Collection

Answer on page 352

1. SCIENTISTS

2. GENERAL IDEAS

3. ACROSS THE SCIENCES

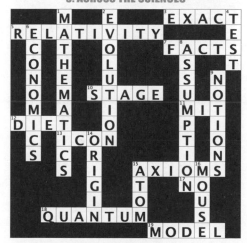

4. ASTRONOMY

5. STUDYING HUMANS

6. BASIC MATHEMATICS

7. APPLIED SCIENCE

CONSTRUCTION
CIVIL
VANE SCALE
ELAID MEDICAL STALL MECHANICAL
ENGINEERING
CREATE STATIC
HYDRAULIC
RAIN MYSELF
LEVER

8. PSEUDOSCIENCE

DATA PROCESS CON
FAKE
FALSEHOODS
MIT IDEA
DEFINES MYTHS
TEN
PREDICT PIRATE
CHANCE
BELIEFS

9. SAVING THE PLANET

ENVIRONMENT
RESPIRE CLIMATE
STATE
WARMING
WIND
TEMPERATE
SOLAR
FOOTPRING

10. GALILEO GALILEI (1564-1662)

CLOCK
ITALIAN
ART JUPITER CHART
MOON
HERESY PLUS
SIMILAR
VENUS
STARS

11. ISAAC NEWTON (1642-1727)

GOAL
PROFESSOR
BRITISH PRISM
CAUSE
SIR
LAWS
APPLE ALCHEMY
LEIBNIZ
PROCESS

12. ALBERT EINSTEIN (1879-1955)

PRINCETON
ENERGY
NIL
ATOMIC COUSIN
USA
GERMANY
VIOLIN
PHYSICS
OIL

304

13. CELESTIAL BODIES

14. FORENSIC SCIENCE

15. HEALTH SCIENCE

16. SPACE EXPLORATION AND OBSERVATION

17. NAVIGATION

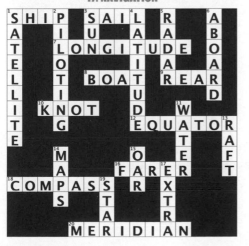

18. GEOGRAPHY

19. FOOD SCIENCE

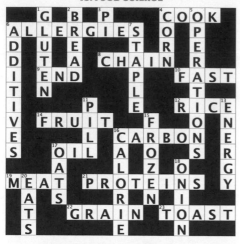

20. THE ENVIRONMENT

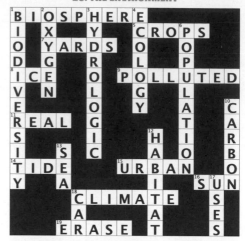

21. ANIMALS AND PLANTS

22. SCIENCE AND FACTS

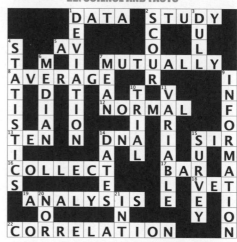

23. MEASUREMENT

24. RECONSTRUCTING THE PAST

25. ARCHIMEDES (C. 287-212)

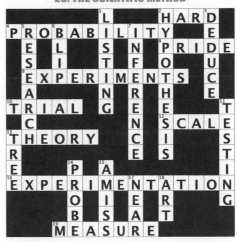

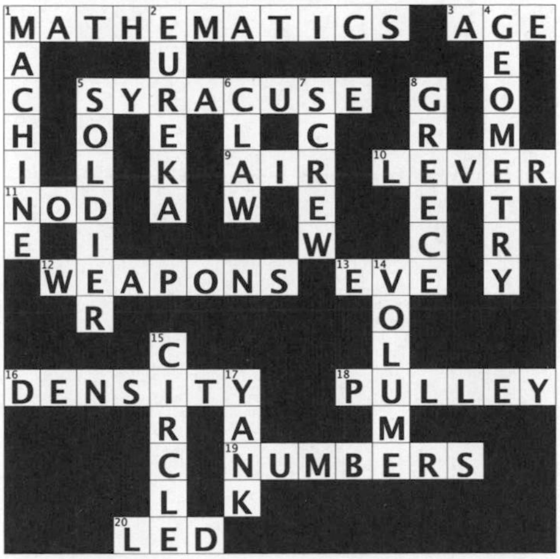

Across/down answers visible: MATHEMATICS, AGE, SYRACUSE, GEOMETRY, AIR, LEVER, WEAPONS, EVE, DENSITY, PULLEY, NUMBERS, LED, MACHINE, CIRCLE

26. THE SCIENTIFIC METHOD

HARD, PROBABILITY, PRIDE, EXPERIMENTS, TRIAL, SCALE, TESTING, THEORY, EXPERIMENTATION, MEASURE

27. PYTHAGORAS (C. 580-500 BCE)

TRIANGLES, HARMONY, PRIME, PLATO, MUSIC, LIGHT, ROT, THEORY, GEOMETRY, ART, PROOF, CRISIS

28. INVENTIONS

TELEPHONE, PILL, WHEEL, NEON, CAR, ROCKETS, SERENDIPITY, TOOLS, FIRE, PRESS, ADA, EDISON, INTERNET, CRAYON

29. MECHANISMS AND TOOLS

GASKETS, MECHANICS, ROTOR, LOCK, TIDAL, GORGE, FASTENERS, ACES, NAV, NAILS, SCREWS, NIL

30. SCIENCE BRANCHES

NASA, SPACE, WEATHER, BOTANY, MOM, ZOOLOGY, PHYSICS, EAT, REF, ARCHEOLOGY, ASTRONOMY, ANT

31. MARIE CURIE (1867-1934)

POLISH · PIERRE · NOBEL · CANCER · RADIOACTIVITY · TEAM · MARIE · DAUGHTER · LAWS · CHEMIST · SEXISM · PARIS · BERATE · NOBEL

32. THE EARTH

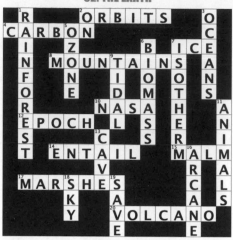

ORBITS · OCEANS · CARBON · ICE · MOUNTAINS · ISOTHERMAL · NASA · EPOCH · ANIMALS · ENTAIL · MAMMALS · MARSHES · VOLCANO · RAINFOREST

33. SCIENCE AND MYTH

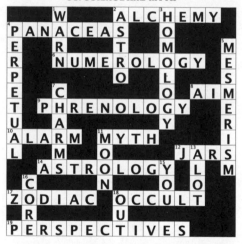

ALCHEMY · PANACEAS · NUMEROLOGY · MESMERISM · PHRENOLOGY · AIM · ALARM · MYTH · JARS · ASTROLOGY · ZODIAC · OCCULT · PERSPECTIVES · PERPETUAL

34. EUCLID (C. 300 BCE)

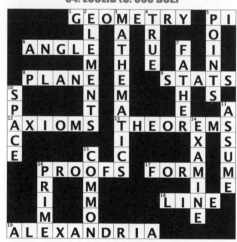

GEOMETRY · PI · POINT · ANGLE · FAN · PLANE · STATS · SPACE · AXIOMS · THEOREMS · ASSUME · PROOFS · FORM · LINE · ALEXANDRIA · MATHEMATICAL

35. SPEED

ACCELERATION · PROOF · COMET · TIME · VECTOR · TEN · RATE · PIE · MILEAGE · TURN · TOPS · CREST

36. TIME

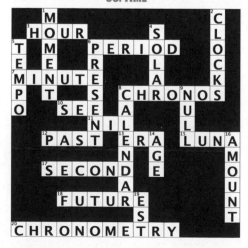

HOUR · CLOCK · PERIOD · TEMPO · MINUTE · CHRONOS · SEE · PAST · ERA · LUNA · SECOND · MOUNT · FUTURE · CHRONOMETRY

37. EXPERIMENTS

38. BIG IDEAS

39. ERATOSTHENES (276-194 BCE)

40. INSTRUMENTS

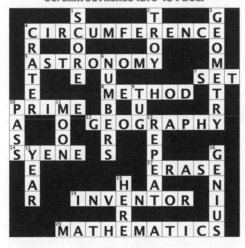

41. AUTOMATION

42. SCIENTIFIC THINKING

43. SYSTEMS

```
N E T W O R K S   C L A S S
  C             R       O
B I O S Y S T E M   R   C I A L
  S         O       E   I
M E   Y     L   A R T I F I C I A L
E   S T     A     V       A
T H E E D     P R E V E N T
O   M             E
D       S T R U C T U R E
  C O M M U N I C A T I O N S
O             I   U     A
P   C A T E G O R I E S   F
E                         S E A
```

44. SCIENCE AND PHILOSOPHY

```
        B   D E M O C R I T U S
    D U     Y         L
N E E D   N A T U R A L       P
  S   D   H         N A S A   A
  C   H   P       L           S
A R I S T O T L E             C
  R   S   A       C           A
  T         L O G I C A L
A S K S             U
L         C O N F U C I U S
I     F             S
G   M E T A P H Y S I C S
N   A
S O C R A T E S   K A N T
```

45. NUMBERS: THE LANGUAGE OF SCIENCE

```
  S   B   R
D E C I M A L   N   I
  X   N   T     E   R
  A   A   I M A G I N A R Y
L O G   O   A   E   A
O   E   N   T   S E T S
O   S   G A P   I       O
P R I M E   L   V   F   N
  M   Q   Z E R O   O   A
  A   U N I T E     R   L
  L   A     N O R M A L
      T   T W O     U
    P I           L
    I       R A T I O
E V E N   C U R V E
```

46. BIG QUESTIONS

```
  M   U   W E A T H E R
  A L O N E   B       E
  R   I     O   H U   I
  S O L V E   V   U   G
      E   D R E A M I N G
      R   R       A
  C O N S C I O U S N E S S
B   M   E   V
A   E     M A T T E R
C A N C E R   O
T       O V E R       B
E     B       T I M E R
R     A   A   A       A
I     I N T E L L I G E N C E
A     N S
```

47. GRAPHICS

```
S E T   G   D O C
H     I R A M   O   V
E     L A P   O   M U S I C
E     L P   D E E P   U
T     U H   E   U   U
S T A S I S L   T   A
  T   T     I   E A R L Y
C H A R T S N E A R   I
  T       Y   G       Z
  I       M   B L U E P R I N T S
  O S O L O L       I
  N     L S   N O T A T I O N
  S       S     A     N
        D I A G R A M
```

48. SCIENCE IN SCHOOL

```
P     H     P L A N
H U M A N I T I E S   R
Y   S           T
S C I E N T I F I C   S O
I   O           O   C   P
C   R           M   O   E
L A B S         P   M A R
B           M   U   I     A
O   M A T H E M A T I C S
T       E   E   T     S O
A   L I T E R A T U R E C I
N   O   X     N   S     A
E W     A     S   I   L
Y       M     T R A I L
C L A S S E S   E   Y
```

310

49. SCIENCE FICTION

50. SCIENCE TIDBITS

51. THE ATOM

52. CHEMISTRY

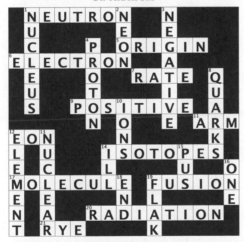

53. GEOLOGY

54. SOLIDS

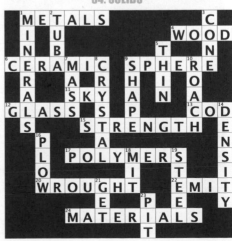

55. GRAVITY

```
E   N   T       F   A
ACCELERATION        A T
R   W   A       U   T R
T   T   I     R E A R A
H   M O M E N T U M   A
    O   N E     S P A C E
  S     E     E R   I   T
  I   F O R C E   I N     M O O N
  G   L   G       S     N
T H E O R Y     S H E   N
  S A W   F   T   F   R
  S       M A G N E T   O   E   A
W E I G H   L     I   R   A
  L       L   N O R M A L
  Y       L           S
```

56. ELEMENTS

```
O   C L E A R       S
X   A       N E O N   D   F
H Y D R O G E N     D     L
G   B   O   I       I     O
E   O   L I T H I U M     U
A N O N   D   R     M   O R E
A T O M   C O R E     O   I
O R A C L E     G     H O N E
M     H   H E L I U M     E
    S I L I C O N
    E   O   L     C A L C I U M
    A I R     M   A
    T I N     N I T R O G E N
    I R O N   U       E
    N   E   M   S U L F U R
```

57. MOLECULES

```
L     U N I T     S A M E
E   C     C   A   I     B
A T O M S   H   F   X   O
D   M   A   E L E C T R O N E
    P   M       U       S
  B O N D S   I O N   R
  U   U   C     R E A L
  N U C L E A R   L   L
S D   T   L     O I L Y
M A S S     R       S
A         I N O R G A N I C
L A S E R S   E     A   H
L   P   A   O     N   Z O O   O
  C O V A L E N T   I     O
    T     T     C A M E L
```

58. CHEMISTRY LABS AND EXPERIMENTS

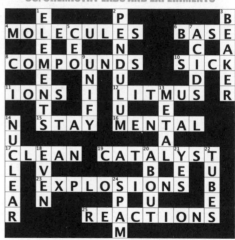

```
    E       P       B
M O L E C U L E S     B A S E
    E     E   N     C   A
C O M P O U N D S   S I C K
    E     N   U     D   E
I O N S   I   L I T M U S   R
    T   I   F U   E
N   S T A Y   M E N T A L
U   I       U       A
C L E A N   C A T A L Y S T
L V       B   E   U
E   E X P L O S I O N S   B
A N         P   U       E
R     R E A C T I O N S
            M
```

59. GASES

```
    N   L I Q U I D
A R G O N   E   O   R
I   B   T   N E A T   D
R   L   D   A       D   O
  H E L I U M   C   O
    O   M E T H A N E
    W A X   O   L
  O   I N N   I
  O   D   A   G R E E N
O Z O N E   S   N   O R E
  E   O   S       T
    N I T R O G E N
F U E L   E   O   E
  O       A   A   T
  F L A M M A B L E
```

60. LIQUIDS

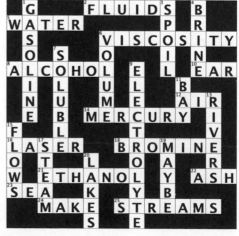

```
  G     F L U I D S   B
W A T E R       P   R
  S       V I S C O S I T Y
  O   S   O       I   N
A L C O H O L   E   L   E A R
I   L   U   L   B
N   U   U   E       A I R
E   B   M E R C U R Y   I
F   L   T       V
L A S E R     B R O M I N E
O   T   L     O   R
W   E T H A N O L   Y   A S H
S E A   K     B
    M A K E   S T R E A M S
          S   E
```

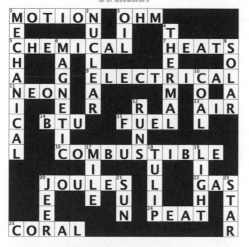

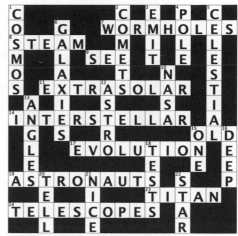

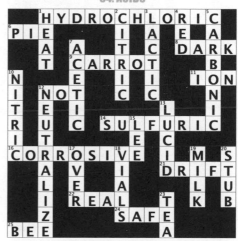

66. OCEANS AND SEAS

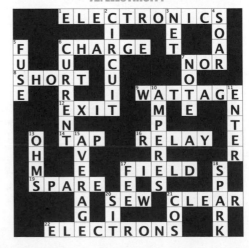

73. OPTICS

74. ROCKETS AND SPACECRAFT

75. TEMPERATURE

76. NICOLAUS COPERNICUS (1473-1543)

77. ENRICO FERMI (1901-1954)

78. NUCLEAR PHYSICS

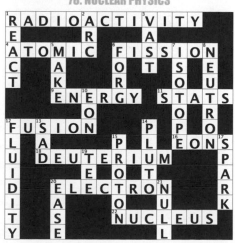

79. THE COSMOS

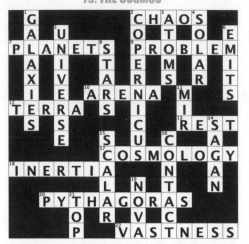

80. THE SUN

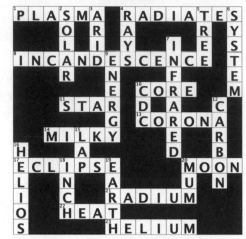

81. CLIMATE

82. ARISTOTLE (384–322 BCE)

83. WATER

84. LIGHT

85. COLOR

P	R	I	S	M	S				
I	N	F					C		H
	R	R					H		U
	G	R	E	E	N		P	U	R	P	L	E
R	A	A	D			P	R		O
O	R	R				I			M			V
S	P	E	C	T	R	U	M		A			I
E		D				N			T		P	I	N	K
		I				E			I			O
W	H	I	T	E			C	O	O	L		O
T		R		A						E		R	A	N	G
I			B	R	O	W	N				T	E	A
N			L				D	U	L	L			N
T			U										G
		Y	E	L	L	O	W		J	A	D	E

86. SHAPES

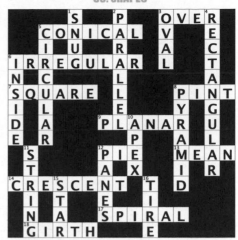

			S		P		O	V	E	R
	C	O	N	I	C	A	L		O	V	A	L
			C		U					R	E	C	T
I	R	R	E	G	U	L	A	R				A
		C			A		L					N
	S	Q	U	A	R	E			P	I	N	T
	D					L		Y			U
	E					P	L	A	N	A	R
						P		R		A	I	D		MEAN
	S					I	E	M		A
	T				P	I	E		M			R
C	R	E	S	C	E	N	T		T	D
	I		T			S		I
	N		A				S	P	I	R	A	L
	G	I	R	T	H			E

87. SURFACES AND TEXTURES

S	P	H	E	R	I	C	A	L		A		V
E	X	T	E	R	I	O	R		M		T	W	I	N
					V	E	I	L			E
	L	E	V	E	L		N		L	I	N	E
	D	X		A		T		A	Y
	R	A	C		E	V	E	N
T	E	X	T	U	R	E
	M					V		S	I	P
S	M	O	O	T	H			U
I	T	I	N	Y		F	A	C	A	D	E
D	E					O		B
E		N	Y			M		L			G	E	O	D	E	S	I	C

88. COMETS

	M	E	T	E	O	R	S		H
T		C		V			A
I	C	E		E		S	M	A	L	L
M		E		R	U	T		R		E
E				S		A		E
D		S	N	O	W	B	A	L	L	S	E	Y	E
U		T			I				A
S		I		T	A	I	L		S
T	S	T	A	R			C	O	B	A	L	T
	L						S
		P	E	R	I	O	D	I	C
R	A	R	E		R		L
O		I		O		V	A	L	U	E
C	O	R	B	I	T	R
K		T		D

89. METEORS

	O		F		I		M		E	V	E	S
I	N	C	A	N	D	E	S	C	E	N	T
	E		L		H		T		A		S	T	A	R	S
	L		S	H	O	W	E	R
	C				O		O
I	R	O	N		T		R
	A				I		I		C
	T				N	O	T	E	O	B
R	O	C	K	S		S	T	R	E	A	K
		S			O		L
		T			C	O	M	E	T
M	E	T	A	L	L	I	C
		E
A	S	T	R	O	N	O	M	Y

90. THE MOON

	E	A	R	T	H		C		E
	S		R		O	B	E	Y	C	L
P	H	A	S	E	S	W		C		L
H		T	A				A	L	D	R	I	N
O		E				E		P
B		L	U	N	A	R		T	P	O	S	E
E		I		O	R	B	I	T
		T		C		D	E
	S	T	A	D	K
M	E	E		I		N	A	S	A
Y					N		I	G	H	T	H	A	L	F
T	I	T	A	N		A			A
H									U	L
S								W	I	L	L

91. MARS

92. PHARMACOLOGY

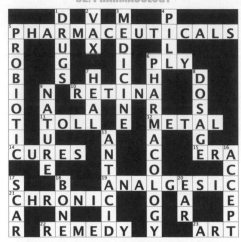

93. FORCES

94. LEONARDO DA VINCI (1452-1519)

95. RATE

96. AVIATION

97. ANGLES

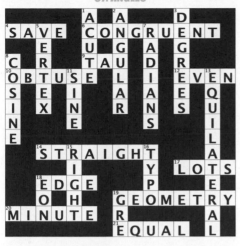

98. AREA

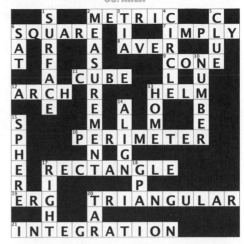

99. VOLUME

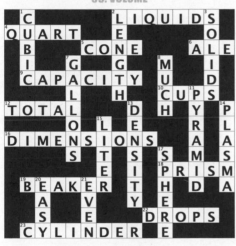

100. ROCKS

101. HOUSEHOLD CHEMICALS

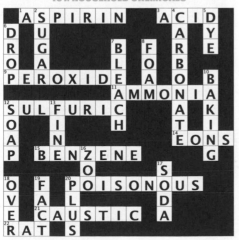

102. BIOLOGY

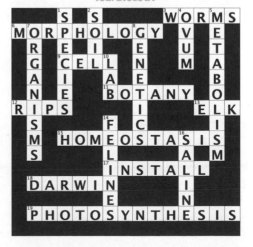

103. EVOLUTION

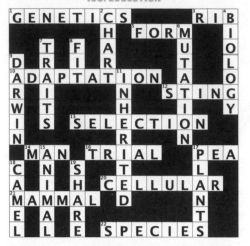

104. GENETICS

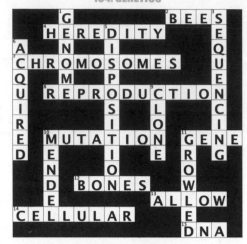

105. BOTANY

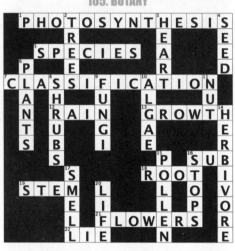

106. ZOOLOGY

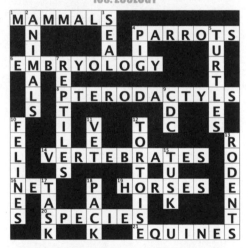

107. THE BRAIN

108. MEDICAL PRACTICE

109. ORGANS

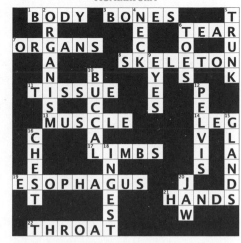

110. ANATOMY

111. DISEASES, ILLNESSES, AND AILMENTS

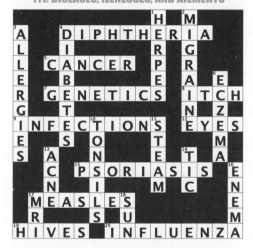

112. OPTOMETRY

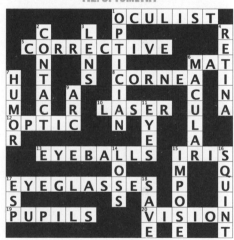

113. CELLS

114. ANIMALS

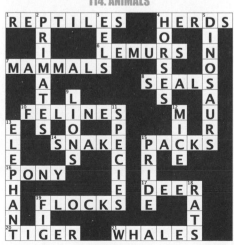

115. INSECTS

```
M E T A M O R P H O S I S P
N     O           P
A N T S   T E R M I T E     I
P   M     H               D
H   O     S P I T E   F L E A
I   L                     R
D     W                   S
  C O C K R O A C H E S
  G           A         A
W       S     S P R A Y I N G
O             P             G
R     T     F             B
M O S Q U I T O   L A R V A E
S     C     U     L         E
  B L A C K T     Y         S
```

116. MORE ABOUT ANIMALS AND INSECTS

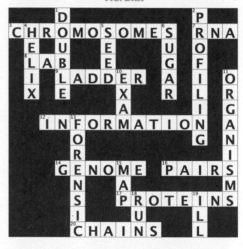

```
B E A R     L I O N     F
S     S     A     F     L
S C O R P I O N   F L I E S
N   R   I   T     I     A
A   C   D       T I C K S
K   B E E T L E S E
E   R   I     S   E
E L     I   G O A T   B E E S
S   A N T   E   S   A   H
  R H I N O S   R O D E N T A
R H I N O S   U         R
  I         C           K
  C R O C O D I L E S   S
```

117. VIRUSES AND BACTERIA

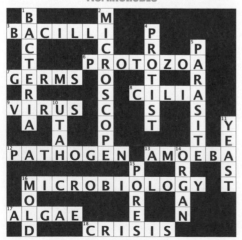

```
M I C R O B E S   T
N         B   S O R E S
F         O   C   R   N
E     H   L   A   A N E W
C O R O N A V I R U S   C
T     S   S   D   S   H
I     T       U       O
O             C O O L E R
I N F L U E N Z A     E
S   E         O       R
A   E     P N E U M O N I A
S Y P H I L I S
  A
  M U M P S
```

118. DNA

```
            D           P
C H R O M O S O M E S U   R N A
E     U     E       S G   O
L A B   E         S U G A R
I X     D L A D D E R   A R
X   E     X       R     I
        I N F O R M A T I O N G
        O             A
G E N O M E   P A I R S     M
N       A             S
S       I P R O T E I N S   L
I       U         U       L L
C H A I N S
```

119. MICROBES

```
  B     M
B A C I L L I   P
  C     I   R       P
  T   P R O T O Z O A
  E   O S   T   A   R
G E R M S   C I L I A
  R   O     S   S   S
V I R U S   T   T   I
  A   T           T Y E
P A T H O G E N   A M O E B A S
      P         R     T
  M I C R O B I O L O G Y
  O       R     A     T
A L G A E R     N
  D     C R I S I S
```

120. RODENTS

```
        P   C           R
C O L O N Y       B     I
R   R   L   H A M S T E R S
C   C   O         A     E
S Q U I R R E L S A V
    P   O         E F E E L
  M I C E         R     E
G   N       M A M M A L S M
O   E   G   A         M
P U S   U   A         I
H       I       T E E T H
E       N               N
  H E R B I V O R O U S  G
S E A   A   L
        U S A
```

121. PLANTS

122. FUNGI

123. MAMMALS

124. BIRDS

125. FISH AND AQUATIC ANIMALS

126. REPTILES

127. THE COMMON COLD

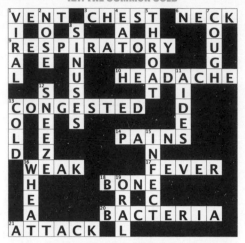

128. THE HUMAN FACE

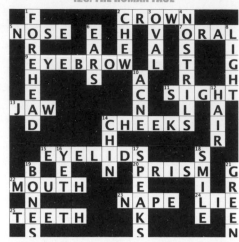

129. CHEMISTRY AGAIN

130. IMMUNITY

131. TOXICOLOGY

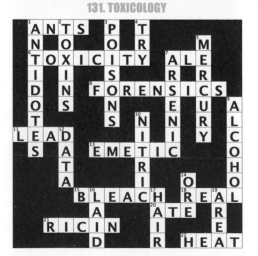

132. COVID-19

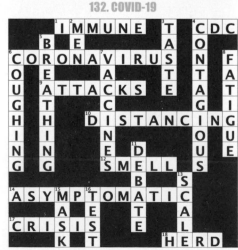

133. CELL BIOLOGY

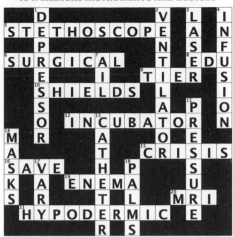

134. MEDICAL INSTRUMENTS AND DEVICES

135. CHARLES DARWIN (1809-1882)

136. AGRICULTURE

137. FORESTRY

138. THE BODY

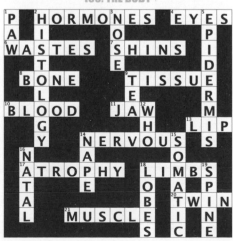

139. THE HEART

140. PARASITES

141. GARDENS

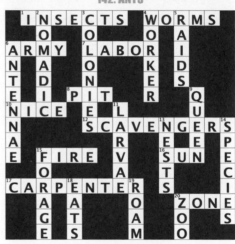

142. ANTS

143. BEES

144. VETERINARY SCIENCE

326

145. SPORTS INJURIES

146. REPRODUCTION

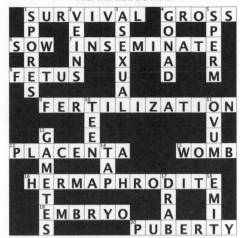

147. FIRE

148. NATURAL SHELTERS

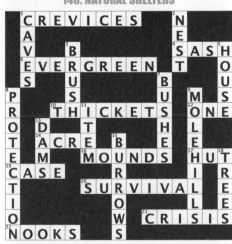

149. MECHANICS

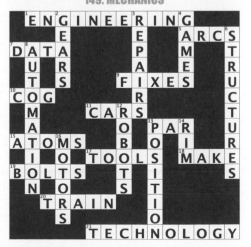

150. WEAPONRY

151. AIRPLANES AND FLIGHT

152. NAVIGATION

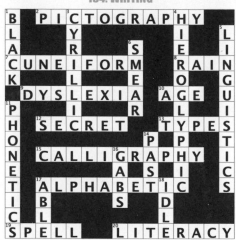

153. ARCHITECTURE

154. WRITING

155. THE WIND

156. POLLUTION

157. HYDRAULICS

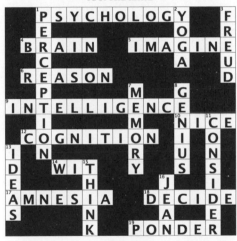

158. LOCOMOTION AND PHYSICAL MOVEMENTS

159. THE MIND

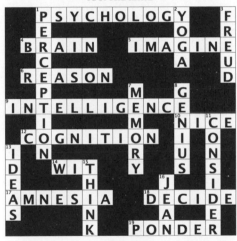

160. PSYCHOLOGY

161. EMOTIONS

162. ARCHEOLOGY

329

163. ANTHROPOLOGY

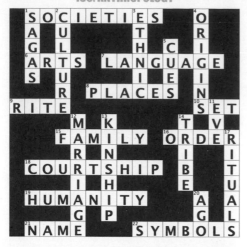

164. SOCIOLOGY

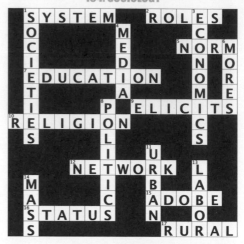

165. LANGUAGE

166. BODY LANGUAGE

167. SIGMUND FREUD (1856-1939)

168. MEMORY

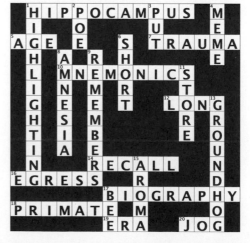

169. INTELLIGENCE

170. LEARNING

171. THE MOUTH

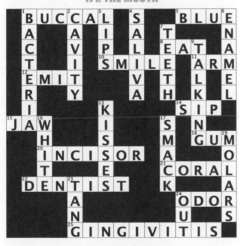

172. VISION

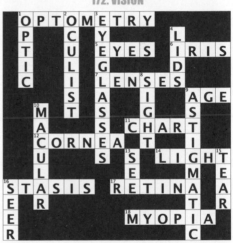

173. SPEECH

174. CRIMINOLOGY

331

175. BUTTERFLIES

176. SLEEP

177. PLANETS

178. ELECTRONICS

179. RESPIRATION

180. CONTINENTS AND PLACES

181. MAPS

182. VOLCANOES

183. RIVERS

184. LAKES

185. THE RAIN

186. ECLIPSES

187. PRIMATES

188. SOIL

189. NUTRITION

190. WILDLIFE

191. VERTEBRATES

192. ARTHROPODS

193. FLOWERS

194. ANIMAL BEHAVIOR

195. CRYSTALS

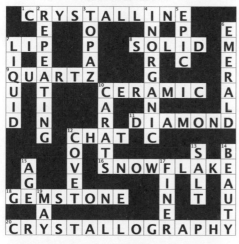

196. THE NORTH POLE

197. THE SOUTH POLE

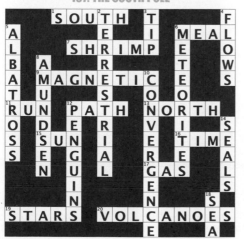

198. KINSHIP

199. MENTAL HEALTH

200. TREES

201. ALGAE

202. DISEASE SPREAD

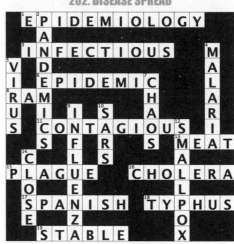

203. AERONAUTICS

204. MARINE BIOLOGY

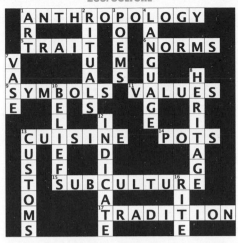

207. POLYMERS

208. GLACIERS

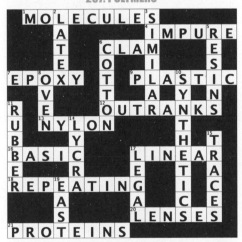

209. ENGINEERING AGAIN

210. HANDS

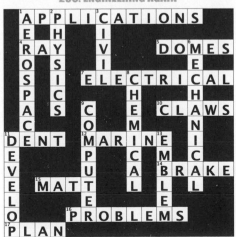

211. CRYPTOGRAPHY

212. EGYPTOLOGY

213. WASTES

214. FUTUROLOGY

215. AGING

216. BLOOD

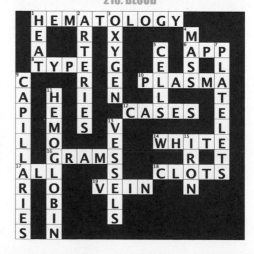

217. SANITATION

218. THE THROAT

219. MUSEUMS

220. CLOUDS

221. SMELL

222. PEDAGOGY

223. BONES

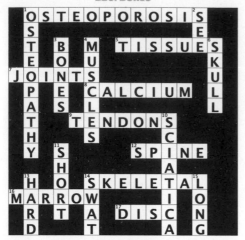

224. FEET

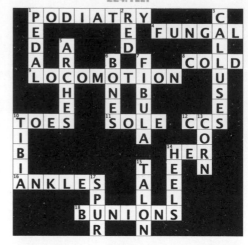

225. EARTHQUAKES

226. SEALS

227. PARAPSYCHOLOGY

228. DISEASES AGAIN

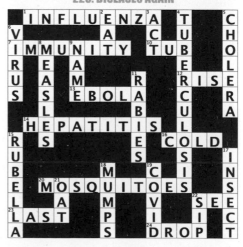

229. POISONS

230. URBANOLOGY

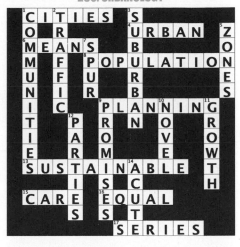

231. WOOD

232. CROPS

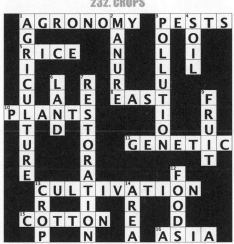

233. SKIN

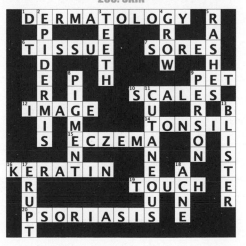

234. THE LIVER

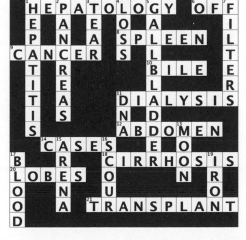

235. MUSCLES

236. FRUITS

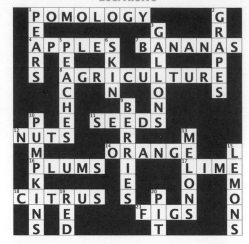

237. GRASS

238. THE EAR

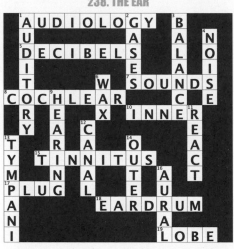

239. PREHISTORY

240. GENETIC ENGINEERING

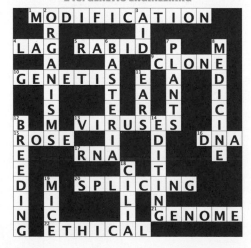

241. DEFEATING CANCER

242. MOSQUITOES

243. BALLISTICS

244. QUANTITY

245. VIBRATIONS

246. MUSICOLOGY

247. MIRRORS

248. VECTORS

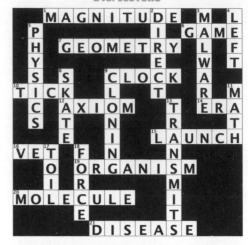

249. SNOW

250. ICE

251. MASS

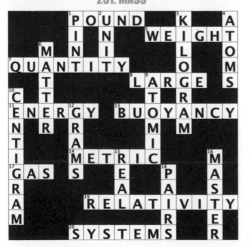

252. DENSITY

253. LASERS

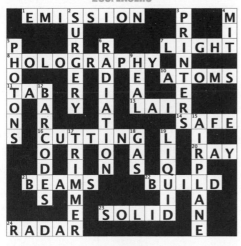

254. THE MILKY WAY

255. METALS

256. SPHERES

257. CARBON

258. HYDROGEN

259. OXYGEN

260. NITROGEN

261. FOOD ADDITIVES

262. ROSALIND FRANKLIN (1920-1958)

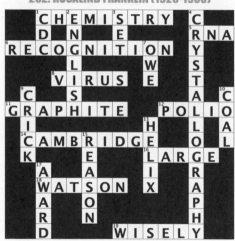

263. GOLD

264. SILVER

265. THE NERVOUS SYSTEM

AUTONOMIC · MOOD
AXON
SIGNALS
PERIPHERAL · FIBERS
NETWORK
NEURONS
NERVE · MIRROR · CIRCUIT
GLIAL · SPINE · WIRE
SYNAPSES

266. EARTHWORMS

INVERTEBRATE · CLAY · EYES
LIMBS · CREEPING · BURROW · LAND · EGG
ASSET · SKIN · LONG · NECK · CROPS · WORM
DRUG · COB · BIT
CYLINDRICAL · SLIME
COMPOST

267. TIDES

CURRENTS · FAR · RIP
FISHING · EBB · NEAP · FALL
FLOODS · SWELL
MOON · STREAM · GRAVITY · GEAR
SIREN · LUNAR · GALILEE
POSITION
NEWTON
PHASE
LEVEL

268. CONSTELLATIONS AGAIN

ANDROMEDA · LYRA · ARIES · SAGITTARIUS
CYGNUS · STARS · BRA
ASTRONOMY
NOT · OTTER · PTOLEMY · LEO · KY
VENUS · VIRGO · VORTEX · LIBRA · BOLTS
TAURUS
ZODIAC

269. MATHEMATICS AND SCIENCE

GALILEO · PLOW · LEG
ARISTOTLE · OVERT
USE · FUNCTION · GEOMETRY · CALCULUS
NEWTON · FORMULA
QUANTITY · EQUALS
QUEEN · PLATO · ASSET
QUARK · ALGEBRA
TRIGONOMETRY

270. THE ANCIENT PAST

PALEOGRAPHY · ARCH · TEXT
PALIMPSEST · ETYMOLOGY · MYRRH · DATING
ASTROLOGY · FOSSILS
AGOG
STEP · DIDDLY · WASTES
SHALL · MANUSCRIPTS
TABLET · SLEEP

271. CARL JUNG (1875-1961)

272. FINGERS

273. LOGIC AND SCIENCE

274. SCIENTIFIC PRACTICE

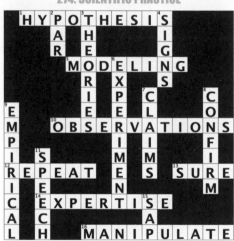

275. FACTS AND FIGURES

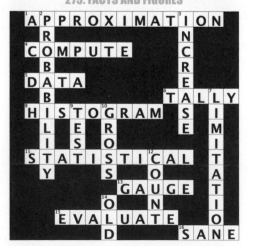

276. COMPUTERS

277. TOOLS, MACHINES, AND INVENTIONS

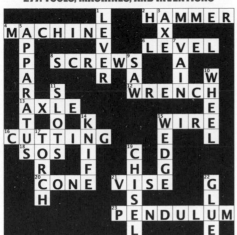

278. DIGITAL TECHNOLOGIES

279. ROBOTS

280. CURRENT TECHNOLOGIES

281. THE RADIO

282. PHOTOGRAPHY

283. THE MASS MEDIA

284. COMMUNICATION

285. TELEVISION

286. SOFTWARE

287. PROGRAMMING

288. ARTIFICIAL INTELLIGENCE

289. THE INTERNET

ONLINE WWW — WIKI — GL D — INTRO — PROTOCOLS — OUT — PROFILE — SHARING — ALL — GOOGLES — MISINFORMATION

290. SOCIAL MEDIA

FACEBOOK FOMO — TWEET NEXT — PROFILE BIO — NICE — MOBILE HE — PLATFORM — INSTAGRAM — TOOL — PLEA — PRIVACY — TIKTOK

291. INFORMATION

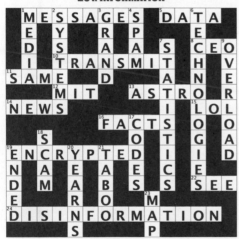

MESSAGES DATA — CEO — TRANSMIT — SAME — MIT ASTRO — NEWS LOL — FACTS LOAD — ENCRYPTED — SEE — DISINFORMATION

292. CYBERSPACE

GIBSON VILLAGE — PECAN PEN — COMPUTER — HOUSE — GLOBAL — CRACK — KITE — IMMERSIVE — ACE — HACKING HORSES

293. WORLD WIDE WEB

WIKIPEDIA WWW — BLOG — BOT — INTERNET ISLE — CEO — FLIP — TRAP SEED — HYPERTEXT — DUE — COOKIE — POT — SERVER CACHE

294. ALGORITHMS

CALCULATIONS — STEPS SYNTAX — SAME PROBLEM — FLOW — OUT NICE — LOGIC RAID — CHAIR — SEARCH — SORTING — FAME

351

295. DATA

296. THE ONLINE WORLD

297. ONLINE PSYCHOLOGY

298. MISINFORMATION

299. MEDIA (NEW AND OLD)

300. SIMULATION

Brimming with creative inspiration, how-to projects, and useful information to enrich your everyday life, Quarto Knows is a favorite destination for those pursuing their interests and passions. Visit our site and dig deeper with our books into your area of interest: Quarto Creates, Quarto Cooks, Quarto Homes, Quarto Lives, Quarto Drives, Quarto Explores, Quarto Gifts, or Quarto Kids.

© 2021 Quarto Publishing Group USA Inc.

First published in 2021 by Chartwell Books,
an imprint of The Quarto Group,
142 West 36th Street, 4th Floor,
New York, NY 10018 USA
T (212) 779-4972 F (212) 779-6058
www.QuartoKnows.com

All rights reserved. No part of this book may be reproduced in any form without written permission of the copyright owners. All images in this book have been reproduced with the knowledge and prior consent of the artists concerned, and no responsibility is accepted by producer, publisher, or printer for any infringement of copyright or otherwise, arising from the contents of this publication. Every effort has been made to ensure that credits accurately comply with information supplied. We apologize for any inaccuracies that may have occurred and will resolve inaccurate or missing information in a subsequent reprinting of the book.

10 9 8 7 6 5 4 3 2 1

Chartwell titles are also available at discount for retail, wholesale, promotional, and bulk purchase. For details, contact the Special Sales Manager by email at specialsales@quarto.com or by mail at The Quarto Group, Attn: Special Sales Manager, 100 Cummings Center Suite 265D, Beverly, MA 01915, USA.

ISBN: 978-0-7858-4010-7

Publisher: Rage Kindelsperger
Creative Director: Laura Drew
Managing Editor: Cara Donaldson
Puzzle Editor: Rebecca Falcon
Cover Design: Beth Middleworth
Interior Design: Danielle Smith-Boldt

Printed in China